Implementing Machine Learning for Finance

A Systematic Approach to Predictive Risk and Performance Analysis for Investment Portfolios

Tshepo Chris Nokeri

Apress®

Implementing Machine Learning for Finance: A Systematic Approach to Predictive Risk and Performance Analysis for Investment Portfolios

Tshepo Chris Nokeri
Pretoria, South Africa

ISBN-13 (pbk): 978-1-4842-7109-4   ISBN-13 (electronic): 978-1-4842-7110-0
https://doi.org/10.1007/978-1-4842-7110-0

Managing Director, Apress Media LLC: Welmoed Spahr
Acquisitions Editor: Celestin Suresh John
Development Editor: Laura Berendson
Coordinating Editor: Aditee Mirashi

Cover designed by eStudioCalamar

Cover image designed by Freepik (www.freepik.com)

Distributed to the book trade worldwide by Springer Science+Business Media New York, 1 New York Plaza, Suite 4600, New York, NY 10004-1562, USA. Phone 1-800-SPRINGER, fax (201) 348-4505, e-mail orders-ny@springer-sbm.com, or visit www.springeronline.com. Apress Media, LLC is a California LLC and the sole member (owner) is Springer Science + Business Media Finance Inc (SSBM Finance Inc). SSBM Finance Inc is a **Delaware** corporation.

For information on translations, please e-mail booktranslations@springernature.com; for reprint, paperback, or audio rights, please e-mail bookpermissions@springernature.com.

Apress titles may be purchased in bulk for academic, corporate, or promotional use. eBook versions and licenses are also available for most titles. For more information, reference our Print and eBook Bulk Sales web page at www.apress.com/bulk-sales.

Any source code or other supplementary material referenced by the author in this book is available to readers on GitHub via the book's product page, located at www.apress.com/978-1-4842-7109-4. For more detailed information, please visit www.apress.com/source-code.

Printed on acid-free paper

I dedicate this book to my family and everyone who merrily played influential roles in my life.

Table of Contents

About the Author

 **Tshepo Chris Nokeri** harnesses big data, advanced analytics, and artificial intelligence to foster innovation and optimize business performance. In his functional work, he has delivered complex solutions to companies in the mining, petroleum, and manufacturing industries. He initially completed a bachelor's degree in information management. Afterward, he graduated with an honor's degree in business science from the University of the Witwatersrand on a TATA Prestigious Scholarship and a Wits Postgraduate Merit Award. The university unanimously awarded him the Oxford University Press Prize. He has authored the book *Data Science Revealed: With Feature Engineering, Data Visualization, Pipeline Development, and Hyperparameter Tuning* (Apress, 2021).

About the Technical Reviewer

 **Anubhav Kesari** is a data scientist and AI researcher by profession and a storyteller by heart. Currently living in Delhi, Anubhav was born and raised in Prayagraj, India, and graduated from the Indian Institute of Information Technology Guwahati with a major in computer science and engineering. His research interests are in machine learning, computer vision, and geospatial data science. He enjoys engaging with the data science community and often giving talks at local meetups as well as for larger audiences. In late 2019, he spoke at PyCon India about hyperparameter optimization in machine learning. He also spoke at PyData Delhi 2019 in a session on machine learning. In his free time, he enjoys exploring nature and listening to classic 90s songs.

Acknowledgments

Writing a single-authored book is demanding, but I received firm support and active encouragement from my family and dear friends. Many heartfelt thanks to the Apress publishing team for all their backing throughout the writing and editing process. Last, humble thanks to all of you reading this; I earnestly hope you find it helpful.

Introduction

Kindly welcome to *Implementing Machine Learning for Finance*. This book is your guide to mastering machine and deep learning applied to practical, real-world investment strategy problems using Python programming. In this book, you will learn how to properly build and evaluate supervised and unsupervised machine learning and deep learning models adequate for partial algorithmic trading and investment portfolio and risk analysis.

To begin with, it prudently introduces pattern recognition and future price forecasting exerting time-series analysis models, like the autoregressive integrated moving average (ARIMA) model, seasonal ARIMA (SARIMA) model, and additive model, and then it carefully covers the least squares model and the long-short term memory (LSTM) model. Also, it covers hidden pattern recognition and market regime prediction applying the Gaussian hidden Markov model. Third, it presents the practical application of the k-means model in stock clustering. Fourth, it establishes the practical application of the prevalent variance-covariance method and empirical simulation method (using Monte Carlo simulation) for value-at-risk estimation. Fifth, it encloses market direction classification using both the logistic classifier and the multilayer perceptron classifier. Lastly, it promptly presents performance and risk analysis for investment portfolios.

I used Anaconda (an open source distribution of Python programming) to prepare the examples. The libraries covered in this book include, but are not limited to, the following:

- Auto ARIMA for time-series analysis
- Prophet for time-series analysis

- HMM Learn for hidden Markov models
- Yahoo Finance for web data scraping
- Pyfolio for investment portfolio and risk analysis
- Pandas for data structures and tools
- Statsmodels for basic statistical computation and modeling
- SciKit-Learn for building and validating key machine learning algorithms
- Keras for high-level frameworks for deep learning
- Pandas MonteCarlo for Monte Carlo simulation
- NumPy for arrays and matrices
- SciPy for integrals, solving differential equations, and optimization
- Matplotlib and Seaborn for popular plots and graphs

This book targets data scientists, machine learning engineers, and business and finance professionals, including retail investors who want to develop systematic approaches to investment portfolio management, risk analysis, and performance analysis, as well as predictive analytics using data science procedures and tools. Prior to exploring the contents of this book, ensure that you understand the basics of statistics, investment strategy, Python programming, and probability theories. Also, install the packages mentioned in the previous list in your environment.

CHAPTER 1

Introduction to Financial Markets and Algorithmic Trading

This is the initial chapter of a book that presents algorithmic trading. This chapter carefully covers the foreign exchange (FX) market and the stock market. It explores how we pair, quote, and exchange official currencies. Subsequently, it covers the stock exchange. In addition, it presents key market participants, principal brokers, liquidity providers, modern technologies, and software platforms that facilitate the exchange of currencies and shares. Furthermore, it looks at the speculative nature of the FX market and stock exchange market and specific aspects of investment risk management. Last, it covers several machine learning methods that we can apply to combat problems in finance.

FX Market

The FX market represents an international market in which investors exchange currency for another. It does not have a principal visible location in which transactions occur, each investor holds their own transaction

T. C. Nokeri, *Implementing Machine Learning for Finance*,
https://doi.org/10.1007/978-1-4842-7110-0_1

records, and each transaction happens electronically. Key market participants self-regulate using guidelines prescribed by a regulatory body within their geographic boundaries.

Exchange Rate

Each official country generally has its own currency. The currency is a class of payments administered and distributed by the central government and dispersed around their geographic boundaries. In relation to foreign trade, an individual or corporation that purchases foreign goods or services and sells them to their local market typically has to exchange currencies. We universally recognize an exchange rate as the ratio of the price of a local currency to a foreign currency. The major currencies include the US dollar ($), euro (€), Great Britain pound (£), Japanese yen (¥), etc. The cross rate is the price of a currency against another, where the US dollar is uninvolved. For instance, euro/GBP is a cross rate between the euro and the sterling.

Exchange Rates Quotation

An exchange rate represents the price of a currency relative to an alternative. We quote currencies directly or indirectly. Using the direct method, the exchange shows how much we have to exchange the local currency for one unit of a foreign currency. For instance, EUR/USD = 1.19. The indirect method shows how much foreign currency trades for one unit of the local currency. For instance, USD/EUR = 0.84.

Exchange Rate Movement

The exchange rate is inconstant; it varies over time. There are several prime factors that influence changes in the exchange rate. For instance, economic and growth factors such as gross domestic product growth

(GDP), inflation rates (consumer price index or GDP deflator), and stocks traded, external debt stocks, current account balance, total reserves, etc. In other instances, rates may react to geopolitical news, natural disasters, labor union activities, social-unrest, corporate scandals, among others. When changes befall, we say that one currency is stronger or weaker than another currency. For instance, with the EUR/USD currency pair, if the euro strengthens, the USD progressively weakens. Let's say EUR/USD opened at 1.2100 and closed at 1.2190; we say that the EUR strengthened since 1 EUR bought more USD at the close than at open.

- Assuming an investor buys 1 million euros at 1.2100 at the open, assuming it will strengthen on that day, but the euro closes at 1.2190, the investor has experienced $7 383. 10(€9000) loss.

$$Euro\ Loss = \frac{+\$1\,000\,0000 = -\$1\,000\,000 =}{-0-} \frac{-€1\,219\,000 + €1\,210\,000}{€9\,000}$$

Bids and Offers

A market maker represents an organization that exchanges currencies on its own account at prices reflected on their systems. Common market makers include banks and brokers. They quote two rates as follows:

- *Bid*: The rate at which market makers buy the currency

- *Offer*: The rate at which market makers sell the base currency

The Left Bid and Right Offer Rule

As tricky as it may be when market makers trade, they are buying and selling at the same time. They buy the base currency on the left side of the quote and sell the currency on the right side of the quote. For instance, if a market maker quotes the EUR/USD at 1.2100/15, they will buy the dollars at €1.2100 and sell them at €1.2115.

EUR/USD	
Bid	**Ask**
1.2100	1.2115

The difference between the bid and offer is called the *spread*. It informs us about liquidity. The more liquid the market, the more narrow the spread. To understand how this works, let's look at the minor currency pairs and major currencies. Currency pairs in emerging markets such as South African rand to rupees (ZAR/INR), Bangladeshi taka to Omani rial (BDT/OMR), among others, have low trading activities and are traded in small quantities. This results in higher spreads when compared to major currencies such as the GBP to the US dollar (USD), Australian dollar to the US dollar, etc. Equation 1-1 shows how we find the spread.

$$Spread = Bid - Ask \qquad \text{(Equation 1-1)}$$

Consider the scenario where the EUR/USD is quoted at 1.2100/15; the spread equals 0.015.

The margin represents the difference between the bid and ask divided by the ask. It can be written mathematically as in Equation 1-2.

$$Margin = (Bid - Ask)/Ask *100\% \qquad \text{(Equation 1-2)}$$

Margin equals 1.2381 percent.

- Assume you are a US tourist who is visiting Europe and wants euros; you must buy the euros using the US dollars you arrived with. The market makers will sell them to you at €1.2115. In contrast, if you are on the seller's side, the market maker buys the currency at €1.2100.

The Interbank Market

The interbank market encompasses a substantial segment of the FX market. It is an international network of large financial companies, especially multinational banks that use their cash balances to trade currencies. We equally recognize the interbank market as the wholesale market. Key participants in this market influence the direction of price movements and interest rate risk through their purchasing activities and sales operations. They set the bid and ask price for a currency pair based on future price predictions. The central bank frequently examines the activities of key market participants to determine the effects of their transactions on economic stability. In addition, they use complex instruments such as fiscal policies and monetary policies to drive price movements.

The Retail Market

The retail market is a narrow segment of the FX market. It encompasses investors who are not directly part of the interbank market. In the retail market, investors carry out transactions over the internet using sophisticated technologies, systems, and software that brokerage companies provide.

Figure 1-1 shows that Electronic Broker Services (EBS) receives prices from banks, brokerage companies, and other financial institutions, and then provides them to retail investors. It provides an electronic platform that enables retail investors to trade with money markers. Key financial institutions in the market include insurance firms, investment firms, hedge funds, etc. EBS compromises leading banks (an alternative to EBS is Thomson Reuters Matching). Established banks include Goldman Sachs, JP Morgans, and HSBC among others. Some of the most popular brokerage firms include Saxo Bank, IG Group, Pepperstone, among others.

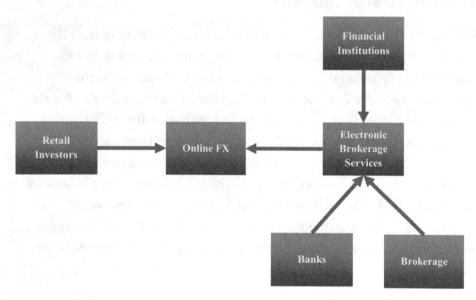

Figure 1-1. *Simple example of the FX market*

Brokerage

A brokerage involves providing a stable platform that facilitates transactions. Brokerage companies typically issue retail investors' trading accounts and the infrastructure for exchanging financial instruments. Figure 1-2 exhibits the various types of brokers.

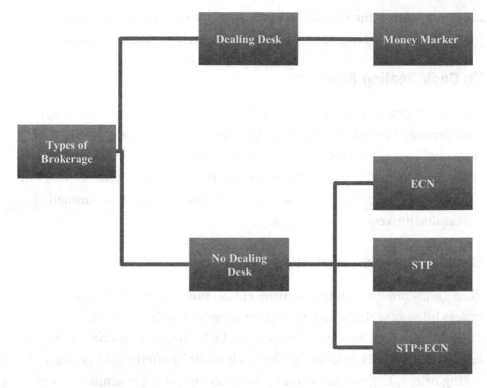

Figure 1-2. *Types of brokers*

Brokerage companies differ subtly in their modus operandi. These companies must be registered and follow compliance standards set by regulatory organizations within their geographic boundaries. Key regulators include the US Commodities and Futures Trading Commission, Australian Securities and Investments Commission (ASIC), and UK Financial Conduct Authority (FCA). Underneath, we consider primary brokerage companies.

Desk Dealing Brokers

Desk dealing (DD) brokers present fixed spreads and liquidity to investors. They establish a market for investors and take the opposing side of investors' orders, meaning that they trade against their customers. They uniformly determine the bid and ask prices based on their own future price predictions. Their rates do not come from the interbank market.

No Desk Dealing Brokers

No desk dealing (NDD) brokers do not pass on investors' orders through desk dealing. They also do not take the other side of the trade executed by investors. To generate revenue, they charge a small commission and/or slightly influence the spread. There are two principal NDD brokers, namely: 1) electronic communications network brokers and 2) straight-through processing brokers.

Electronic Communications Network Brokers

Electronic communications network (ECN) brokers ensure investors' orders interact with the orders of other investors in the network. Traditionally, investors include commercial banks, institutional investors, hedge funds, etc. They trade against each other by offering bid prices and asking prices. To generate revenue, they call for a small commission and attract an enormous pool of investors through robust marketing initiatives.

Straight-Through Processing Brokers

Straight-through processing (STP) brokers direct investors' orders to liquidity providers with access to the interbank market. At most, these brokers favorably get many bid prices and ask prices from many liquidity providers such as Citibank, Barclays Bank, Morgan Stanley, among others, who carefully sort them and then offer investors a price with a markup. To generate revenue, they call for a hefty commission. Unlike the two types of brokerage firms mentioned earlier, these brokers are not concerned with influencing trading activities.

Understanding Leverage and Margin

The FX market offers more excessive leverage compared to the stock market. Leverage represents the amount that brokers loan investors to transact. It exposes an investor to a position that they would not have with their cash balance. Brokers offer trading accounts that use leverage through margin trading. The margin represents the difference between the total value of the positions held and the total value of the loan. Margin quantity increases as a result of a decrease in leverage or vice versa. Leverage enables investors to execute numerous trades they would ordinarily trade using the current account; you can basically look at it as credit without the requirement of collateral. It is often depicted in a ratio. Standard trading accounts have leverage types that range from 1:10 to 1:100. However, some brokerages do offer accounts with leverage up to 1:1000. The lower the ratio, the higher the capital needed to execute a trade. Investors exert leverage to generate more returns with slight price changes. For instance, an account with 1:1000 enables investors to execute numerous trades compared to 1:10. Also, the higher leveraged accounts amplify profits and magnify losses. To some extent, an account's leverage type gives a slight idea of an investors' risk profile. For instance, an investor expecting to rapidly generate profits in the market will have a high leveraged account, and a more conservative investor will have a low leveraged account.

The Contract for Difference Trading

The contract for difference (CFD)[1] represents a financial derivative with a value that comes from a financial asset. It enables investors to generate profits from price differences rather than safeguarding an asset.

---

[1]https://web.archive.org/web/20160423094214/https://www.nsfx.com/about-nsfx/risk-disclosure/

Consequently, investors can speculate on price movements. Besides CFDs, there is a more complex financial derivative known as *futures*. Unlike CFDs, which are privately traded through brokers, futures are traded through a large exchange. With futures, an investor on the buyer's side is obliged to execute the trade when the contract expires, and an investor on the seller's side ought to deliver the asset at a certain period. Futures have an expiry date, and there is a cap on the number of trades that an investor can execute at a certain period. Given that, the future has stricter regulatory mechanisms than CFDs.

The Share Market

We equally recognize the share market as the equity market or the stock market. It is also one of the most liquid markets in the world. It is an international market mostly made up of large financial companies that trade listed stocks. A share represents the ownership of equity. A company sells equity to raise capital.

Raising Capital

At most, companies seek to expand their business operations, but funds are constrained by capital. They may privately or publicly seek debt financing or equity financing. Debt financing involves borrowing funds using collateral. The most conventional source of debt financing is big commercial banks. Most startups cannot source financial capital from these banks because they are at high risk (most startups fail). They use alternative debt financing sources like micro-lenders, angel investors, and seed investors depending on the phase they are at in their entrepreneurship journey. More established companies may raise capital by selling their shares to the public.

Public Listing

If an established company wants to gain a large financial capital, then it can list on a public stock exchange.[2] Before they exchange company stocks, they must first comply with the accreditation process of a commission in a specific jurisdiction and make an initial public offering.

Stock Exchange

The stock market exchange facilitates the exchange of shares between companies and investors. The most popular stock exchanges include the London Stock Exchange, New York Stock Exchange, and NASDAQ among others. Figure 1-3 shows a simple example of the stock exchange market.

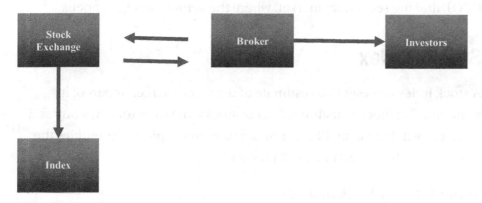

Figure 1-3. *A simple example of the stock exchange market*

Figure 1-3 shows that the brokers interact with the stock exchanges and pass prices to investors. Also, the stock exchange releases a stock market index. There are two markets on the stock exchange. First, there is the *primary market*, where a company shows intent of enlisting in the stock exchange by delivering an official press release, partaking

---

[2]https://www.investopedia.com/terms/i/ipo.asp

in roadshows to make their intentions known to the public, filing for enlistment, and making an initial public offer (IPO), which involves issuing shares prior to enlistment. Post enlistment on the exchange, a company enters the *secondary market* by making a follow-on offer and opening trading activities to the public. This occurs upon satisfying the compliance requirements of the exchange.

Share Trading

The stock market exchange facilitates the exchange of securities like shares, bonds, and stocks. Key markets in the stock exchange include the following: the primary market, where a company may issue shares before they are listed on an exchange, which is known as an *initial public offer* (IPO), and the secondary market, where the actual trading happens.

Stocks Index

A stock index represents an estimate of the stock market or part of its segment. The index constitutes a set of stocks that investors may buy as a coherent whole or in an EFT or mutual fund that typically resembles the index. Table 1-1 shows key stock indexes.

Table 1-1. *Key Stock Indices*

Name	Description
Standard & Poor (S&P) 500	Measures the stock performance of 500 large companies in the United States
Dow Jones Industrial Average (DJI 30)	Measures the stock performance of 30 large companies in the United States
NASDAQ Composite Index	Measures the stock performance of almost all the NASDAQ stock market

The stock major index (S&P 500, DJI 30, and NASDAQ 100, etc.) applies a divisor (often discrete) to divide total market capitalization and get the index value. They compromise stocks across diverse industries. They do not differ from other asset classes by the way we exchange them.

Speculative Nature of the Market

In the exchange market, investors speculate on future prices of an asset and execute trades, so they yield reasonable returns as the price moves toward their speculation. If the price moves in the opposite direction, then investors incur losses. As a result of leverage, investors are exposed to high risk since investors are given room to trade asset classes they cannot afford, meaning they may lose an enormous amount of their capital with slight changes in the market (see https://www.capitalindex.com/bs/eng/pages/trading-guides/margin-and-leverage-explained). Most retail investors trading CFDs lose their capital. An investor must understand the risks associated with trading prior to investing their capital.

Techniques for Speculating Market Movement

Investors use either the subjective method or the objective method to speculate the market or a combination of both. When using subjective means, an investor uses their rationalized belief, experience, opinions of others, and emotions to decide on whether to buy or sell at a specific price. When using objectivity, an investor applies mathematical models to identify patterns in the data and forecast future prices and then decides on whether to buy or sell a currency pair at a specific price. This book only covers machine learning and deep learning models.

Investment Strategy Management Process

Figure 1-4 shows a simple investment strategy management process.

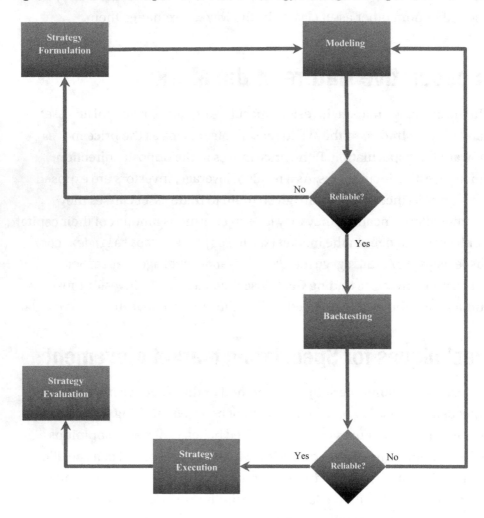

Figure 1-4. *A simple investment strategy management process*

Strategy Formulation

Strategy formulation is the first step in investment management. It involves the following tasks:

- Identifying risk and opportunity

- Setting short-term and long-term objectives

- Identifying resources and ways of organizing, managing, and directing those resources (human, financial, and technical)

- Establishing a structure and policies

Systematic investors model data to draw meaningful insights that influence strategy.

Modeling

After deciding on ways to manage an investment portfolio, the subsequent step involves modeling. It involves the use of quantitative methods. This book focuses exclusively on machine learning and deep learning models from a finance perspective. The subsequent section discusses learning methods applicable to finance, especially investment management.

Supervised Learning

In supervised learning, a model predicts future values of a dependent variable using a function that operates on a set of independent variables based on labels we provide them in the training process. Supervised learning requires us to split data into training and test data (at times validation data too). We present a model with a set of correct answers and allow it to predict unseen answers. There are three primary types of supervised learning methods, namely, the parametric method, the nonparametric method, and the ensemble method. See Figure 1-5.

15

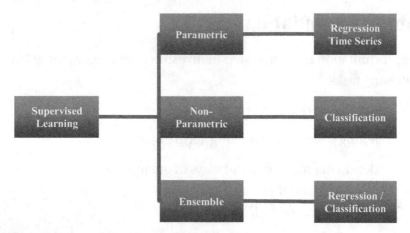

Figure 1-5. *Supervised machine learning*

The Parametric Method

The parametric method is also called the linear method. It makes strong assumptions about the structure of the data. We assume the underlying structure of the data is linear and normal. It handles dependent variables as a continuous dependent variable (a dependent variable that is limited to a specific range). This covers time-series analysis in Chapter 2 and the ordinary least squares model in Chapter 6.

The Nonparametric Method

Unlike the parametric method, the nonparametric method does not have substantial assumptions of linearity and normality. It handles a categorical dependent variable (a dependent variable that is limited to a specific range). There are two primary nonparametric methods: binary classification and multiclass classification.

Binary Classification

In binary classification, the independent variable produces two classes such as no and yes or fail and pass. We code the classes as 0 and 1 and train binary classifiers to predict subsequent classes.

Multiclass Classification

We use multiclass classification when the dependent variable is over two classes such as negative, neutral, or positive. We code the classes as 0, 1, 2...n. The coded values should not exceed 10. The most popular multiclass classification models include random forest and linear discriminant analysis, among others. This book does not cover multiclass classification models.

The Ensemble Method

The ensemble method encompasses both the parametric method and the nonparametric method. We use it when the dependent variable is a continuous variable or categorical variable. It addresses linear regression and classification problems. The most popular ensemble methods include support vector machine and random forest tree, among others. This book does not cover ensemble models.

Unsupervised Learning

Unsupervised learning does not call for data to split the data into training data, test data, and validation data. We do not hand out correct answers to a model; we allow it to form intelligent guesstimates on its own. Cluster analysis is the most prevalent unsupervised learning method. See Figure 1-6.

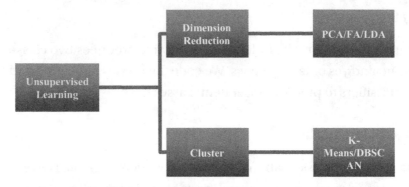

Figure 1-6. *Unsupervised learning*

Dimension Reduction

Dimension reduction is a technique to summarize data by reducing it to small dimensions. We mainly use this technique for variable selection. Popular dimension reduction techniques include the principal components analysis (PCA), which identifies components that account for most of the variation in the data, and factor analysis, which identifies latent factors that account for most of the variation in the data. This book covers dimension reduction in Chapter 5.

Cluster Analysis

Cluster analysis involves grouping data based on similar similarities. It is useful when we have no assumptions about the structure of the data. In cluster analysis, there is no actual dependent variable. The most common cluster model is the K-Means; it partitions the data into k (clusters) with the nearest mean (centroids); it then finds the distance between subgroups to produce a cluster. This book covers K-Means in Chapter 5.

Backtesting

After we develop a model, we must determine how reliable the model is. Backtesting[3] sits between strategy formulation and execution. It involves determining the extent to which a model performs. At most, systematic investors' backtest simulated markets. The easiest way to understand market patterns involves visualizing historical trade executions and price movements. Key Python frameworks that support backtesting include PyAlgoTrade and Zipline.

Strategy Implementation

After finding a reliable investment strategy, we may deploy a model to buy and sell asset classes, thus risking the capital of the investment portfolio. A system may trade manually or be automated, using reliable systematic applications. Key Python frameworks that support paper and live trading include QuantConnect, Quantopia, Zipline, etc. This book does not cover backtesting and live trading frameworks.

Strategy Evaluation

Strategy evaluation involves assessing how well a strategy performs. It enables investors to devise action plans for performance improvements. When analyzing the performance of a strategy, investors mainly focus on the value at risk, the annual rate of return, the cumulative rate of the return value, and drawdown. Investors use these statistics to revise their strategy. This book covers investment risk analysis and performance using Pyfolio.

---

[3]Backtesting Systematic Trading Strategies in Python: Considerations and Open Source Frameworks: https://www.quantstart.com/articles/backtesting-systematic-trading-strategies-in-python-considerations-and-open-source-frameworks/

Algorithmic Trading

Instead of vigorously observing action prices and physically executing orders, an investor may automate or partially automate tasks using sophisticated applications operating a set of predetermined rules (algorithms). This trading technique helps meaningfully reduce redundant tasks, consequently allowing investors to focus on more important duties. Using automated programs eliminates subjectivity. Meaning, investors do not execute orders ostensibly based on some opinion, feeling, or emotion. Instead, they deploy scalable machine learning and deep learning models. This book conceals the art and science of developing and testing scalable machine learning models and deep learning models. High-frequency trading[4] goes hand in hand with algorithmic trading; however, we do not cover the topic in this book. You can apply the models discussed in this book to solve complex problems outside the realm of finance.

This book does not provide any financial advice. It is a technical book that introduces data scientists, machine learning engineers, business, and finance professionals to algorithmic trading by exploring several supervised learning models and unsupervised learning models.

---

[4]High-Frequency Trading: An Innovative Solution to Address Key Issues (harvard.edu): https://corpgov.law.harvard.edu/2014/09/17/ high-frequency-trading-an-innovative-solution-to-address-key-issues/

CHAPTER 2

Forecasting Using ARIMA, SARIMA, and the Additive Model

Time-series analysis is a method for explaining sequential problems. It is convenient when a continuous variable is time-dependent. In finance, we frequently use it to discover consistent patterns in the market data and forecast future prices. This chapter offers a comprehensive introduction to time-series analysis. It first covers ways of finding stationary in series data using the augmented Dickey-Fuller (ADF) test and testing for white noise and autocorrelation. Second, it reveals techniques of succinctly summarizing the patterns in time-series data using smoothening, such as the moving average technique and exponential technique. Third, it properly covers the estimation of rates of return on investment. Last, it covers hyperparameters optimization and model development and evaluation. This chapter enables you to design, develop, and test time-series analysis models like the autoregressive integrated moving average (ARIMA) model, seasonal ARIMA (SARIMA) model, and additive model, to identify patterns in currency pairs and forecast future prices. In this chapter, we use `pandas_datareader` to scrape financial data from Yahoo Finance, and we use `conda install -c anaconda pandas-datareader`.

© Tshepo Chris Nokeri 2021
T. C. Nokeri, *Implementing Machine Learning for Finance*,
https://doi.org/10.1007/978-1-4842-7110-0_2

For time-series modeling, we use the `statsmodels` library, which is pre-installed in the Python environment. We also use `pmdarima`, which is an extension of `statsmodels`. To install it in the Python environment, we use `pip install pmdarima`; in the conda environment, we use `conda install -c saravji pmdarima`. Lastly, we use FB Prophet for high-quality time-series analysis. To install it in the Python environment, we use `pip install fbprophet`; in the conda environment, we use `conda install -c conda-forge fbprophet`. Before you install `fbprophet`, ensure that you first install `pystan`. To install `pystan`, we use `conda install -c conda-forge pystan`.

Anaconda is the most popular open source Python distribution and enables one to manage, install, update, and manage packages (download the platform from `https://www.anaconda.com/products/individual`). You can install the platform on Windows, macOS, and Linux operating systems. Find out more about system requirements and hardware requirements at `https://docs.anaconda.com/anaconda-enterprise/system-requirements/`.

Time Series in Action

Time-series analysis is suitable for estimating a continuous variable that is time-dependent. In this chapter, we use it to identify the structure of sequential data. It is a seamless method for identifying patterns and forecasting future prices of currency pairs. Market data is often sequential and has some stochastic elements, meaning there is an underlying random process. We analyze the historical data of one of the most traded currency pairs in the world, the US dollar ($) and Japanese yen (¥), or the (USD/JPY) pair. We are interested in uncovering patterns in the adjusted closing price of the currency pair across time and then making reliable predictions of price movements. To create a time-series model, first launch Jupyter Notebook and create a new notebook.

Listing 2-1 collects the price data of the USD/JPY pair from November 1, 2010, to November 2, 2020 (see Table 2-1).

Listing 2-1. Scrap Data

```
import pandas as pd
from pandas_datareader import data
start_date = '2010-11-01'
end_date = '2020-11-01'
ticker = 'usdjpy=x'
df = data.get_data_yahoo(ticker, start_date, end_date)
df.head()
```

Table 2-1. *Dataset*

Date	High	Low	Open	Close	Volume	Adj Close
2010-11-01	81.111000	80.320000	80.572998	80.405998	0.0	80.405998
2010-11-02	80.936996	80.480003	80.510002	80.558998	0.0	80.558998
2010-11-03	81.467003	80.589996	80.655998	80.667999	0.0	80.667999
2010-11-04	81.199997	80.587997	81.057999	81.050003	0.0	81.050003
2010-11-05	81.430000	80.619003	80.769997	80.776001	0.0	80.776001

As mentioned, we are interested in the adjusted closing price (Adj Close). Listing 2-2 deletes columns that we will not make use of.

Listing 2-2. Delete Columns and Drop Missing Values

```
del df["Open"]
del df["High"]
del df["Low"]
del df["Close"]
del df["Volume"]
df = df.dropna()
df.info()
<class 'pandas.core.frame.DataFrame'>
DatetimeIndex: 2606 entries, 2010-11-01 to 2020-11-02
Data columns (total 1 columns):
 #   Column     Non-Null Count   Dtype
---  ------     --------------   -----
 0   Adj Close  2606 non-null    float64
dtypes: float64(1)
memory usage: 40.7 KB
```

In this listing, we deleted most columns and dropped missing values. The remaining column is the adjusted closing price column; the format is Adj Close.

Split Data into Training and Test Data

There are 2,606 data points in the time-series data. Listing 2-3 splits the data using the 80/20 split rule (the first 2,085 data points are for training the model, and the remaining are for testing the model).

Listing 2-3. Split Data into Training and Test Data

```
train = df[:2085]
test = df[2085:]
```

Test for White Noise

If time-series data is stationary, then it contains white noise. The most straightforward way to investigate white noise involves generating random data and finding out whether there is white noise in the arbitrary data. Listing 2-4 returns random numbers and plots the autocorrelation across different lags (see Figure 2-1).

Listing 2-4. White Noise Test

```python
from pandas.plotting import autocorrelation_plot
import matplotlib.pyplot as plt
import numpy as np
randval = np.random.randn(1000)
autocorrelation_plot(randval)
plt.show()
```

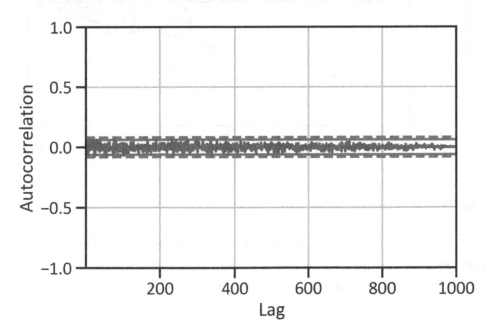

Figure 2-1. *Random white noise test*

Figure 2-1 shows that there is no white noise in the series because there are significant spikes above 95 percent and 99 percent confidence intervals. Listing 2-5 plots training data autocorrelation to show white noise (see Figure 2-2).

Listing 2-5. Training Data White Noise Test

```
autocorrelation_plot(train["Adj Close"])
plt.show()
```

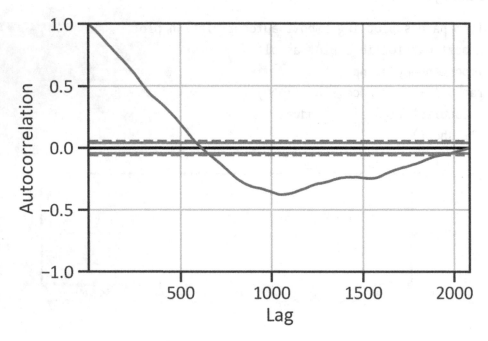

Figure 2-2. *Training data white noise test*

Figure 2-2 indicates for lag 1 there is a sharp decline, and after the 500th lag the line goes lower and approaches zero. Since all autocorrelations are not equal to zero, we can confirm that there is no white noise.

Test for Stationary

The presence of the stochastic (or random) process in ordered data may affect the conclusions. Listing 2-6 checks whether the series is stationary using a unit root test called the augmented Dickey-Fuller test (see Table 2-2). A series is stationary when the mean value of the series data is zero, which means the observations do not vary across time. With the augmented Dickey-Fuller (ADF) test when the p-value is greater than 0.05, we do not reject the hypothesis.

The hypothesis of an ADF test is written as follows:

> *Null hypothesis:* There is no unit root.

> *Alternative hypothesis:* There is a unit root.

A series is nonstationary when the ADF F% statistics is below zero and the p-value is less than 0.05.

Listing 2-6. Augmented Dickey-Fuller Test

```
from statsmodels.tsa.stattools import adfuller
adfullerreport = adfuller(train["Adj Close"])
adfullerreportdata = pd.DataFrame(adfullerreport[0:4],
                           columns = ["Values"],
                           index=["ADF F% statistics",
                                  "P-value",
                                  "No. of lags used",
                                  "No. of observations"])

adfullerreportdata
```

Table 2-2. *F Statistics*

	Values
ADF F% statistics	-1.267857
P-value	0.643747
No. of lags used	6.000000
No. of observations	2078.000000

Table 2-2 highlights that the F statistics result is negative and the p-value is greater than 0.05. We do not reject the null hypothesis; the series is nonstationary. This entails that the series requires differencing.

Autocorrelation Function

Listing 2-7 determines the serial correlation between y and y_t (y_t entails that the observation *y* is measured in time period $_t$). We use the autocorrelation function to measure the degree to which present values of a series are related to preceding values when we consider the trend, seasonality, cyclic, and residual components.

Figure 2-3 shows that most spikes are not statistically significant. In addition, we use the partial autocorrelation function (PACF) to further examine the partial serial correlation between lags.

Listing 2-7. Autocorrelation

```
from statsmodels.graphics.tsaplots import plot_acf
plot_acf(train["Adj Close"])
plt.xlabel("Lag")
plt.ylabel("ACF")
plt.show()
```

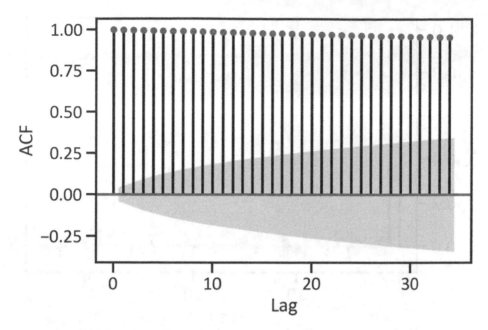

Figure 2-3. *Autocorrelation*

Partial Autocorrelation Function

The PACF plot expresses the partial correlation coefficients not described at low-level lags. Listing 2-8 constructs the PACF plot (see Figure 2-4).

Listing 2-8. Partial Autocorrelation

```
from statsmodels.graphics.tsaplots import plot_pacf
plot_pacf(train["Adj Close"])
plt.xlabel("Lag")
plt.ylabel("ACF")
plt.show()
```

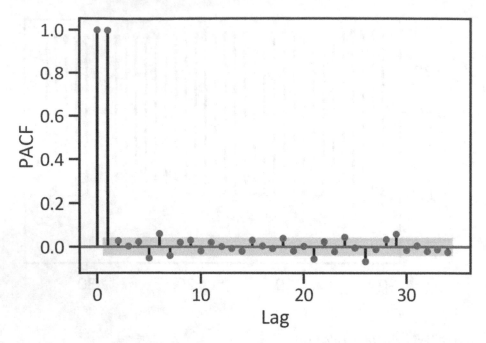

Figure 2-4. *Partial autocorrelation*

We use a correlogram to discover lags that explain the effect outside the 95 percent confidence boundary. For instance, Figure 2-3 shows that there is a significant spike in lag 1 and lag 2. This means the lags can explain all higher-order autocorrelation (at most the second lag is the highest-order lag). It shows spikes that are not statistically significant. (There is a strong positive correlation until lag 2; after lag 2, the p-value is less than 0.05.) The autocorrelation is close to zero, which is around the statistical control (see the blue boundary in Figure 2-3). There is a strong dependency on the time-series data.

The Moving Average Smoothing Technique

The most familiar smoothing technique is the moving average (MA) technique; it returns the means of weight average of preceding and new data points, where weighting depends on the cohesion of the time-series data. In this example, we use Pandas to perform rolling window calculations. After that, we discovered the rolling mean with two fixed-size moving windows. A moving window represents the number of data points for calculating the statistics. Listing 2-9 smooths the time-series data using a 10-day rolling window and a 50-day rolling window (see Figure 2-5).

Listing 2-9. Time Series (10-Day and 50-Day Moving Average)

```
MA10 = train["Adj Close"].rolling(window=10).mean()
MA50 = train["Adj Close"].rolling(window=50).mean()
df.plot(kind="line",color="navy")
MA10.plot(kind="line",color="green",label="10D MA")
MA50.plot(kind="line",color="orange",label="50D MA")
plt.xlabel("Date")
plt.xticks(rotation=45)
plt.ylabel("Adj Close")
plt.legend()
plt.show()
```

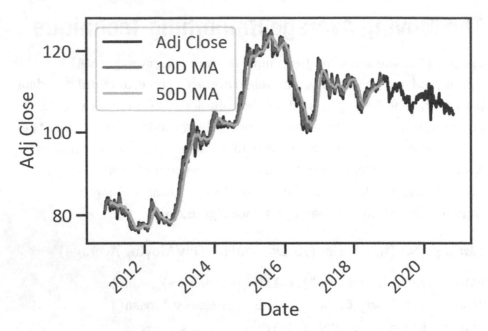

Figure 2-5. *Time series (10-day and 50-day moving average)*

The Exponential Smoothing Technique

The exponential smoothing method weighs values outside the window to zero, where enormous weighted values rapidly die out and minor weighted values gradually vanish. Listing 2-10 smooths the time-series data using the exponential smoothing technique and sets the half-life to 3. Half-life is a parameter that specifies the lag at which exponential weights decay by half. In Listing 2-10, we specified the parameter as 3 since there is a strong positive correlation until lag 2; after lag 2, the p-value is less than 0.05 (refer to Figure 2-3).

Listing 2-10. Develop Smooth Series (Exponential)

```
Exp = train["Adj Close"].ewm(halflife=30).mean()
df.plot(kind="line", color="navy")
Exp.plot(kind="line", color="red", label="Half Life")
```

```
plt.xlabel("Time")
plt.ylabel("Adj Close")
plt.xticks(rotation=45)
plt.legend()
plt.show()
```

Figure 2-6 shows the core structure of the time-series data using the moving average technique and exponential technique.

Figure 2-6. *Time series (exponential)*

Rate of Return

Listing 2-11 estimates and plots the rate at which an asset yield returns annually (see Figure 2-7).

Listing 2-11. Rate of Return

```
pr = df.pct_change()
pr_plus_one = pr.add(1)
cumulative_return = pr_plus_one.cumprod().sub(1)
fig, ax = plt.subplots()
cummulative_return = cumulative_return.mul(100)
cummulative_return_max = cummulative_return.max()
cummulative_return_min = cummulative_return.min()
cummulative_return.plot(ax=ax, color="purple")
plt.axhline(y=cummulative_return_max.item(), color="green",
            label="Max returns: " + str(round(cummulative_
            return_max.item(),2)) + " %")
plt.axhline(y=cummulative_return_min.item(), color="red",
            label="Min returns: " + str(round(cummulative_
            return_min.item(),2)) + " %")

plt.xlabel("Date")
plt.ylabel("Return (%)")
plt.legend(loc="best")
plt.xticks(rotation=45)
plt.show()
```

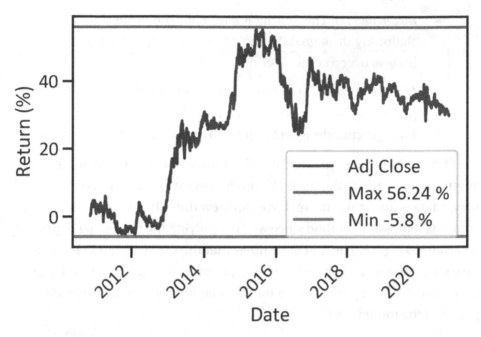

Figure 2-7. *Rate of return*

Figure 2-7 shows a reasonable rate of returns over a period of ten years (from 2010 to 2020). The minimum rate is -5.8 percent, and the maximum rate is 56.24 percent.

The ARIMA Model

In the next section, we use ARIMA to model the series and forecast its future instances. ARIMA is by far the most widespread univariate time-series analysis model. Let's break it down:

- *Autoregressive (AR)*: Linear combination of previous errors. AR considers observation terms of previous observations, including random white noise and preceding random white noise.

- *Integrated (I)*: Transformation to make the series stationary through differencing (estimating the change in rows over a certain period).

- *Moving average (MA)*: Linear combination of previously weighted means (refer to the moving average smoothing techniques covered previously).

Similar to the least-squares model, ARIMA makes strong assumptions about the structure of the data. The model assumes the structure of the series data is linear and normal. We can view the ARIMA model as a complex regression method since we are applying the regressing regress lag1, lag2 to lag = k. The model assumes that the series is stationary. In a case where the series is not stationary, the series must show a trend over time. Also, one can perform data transformation to improve the predictive power of the model.

ARIMA Hyperparameter Optimization

Listing 2-11 finds the best hyperparameters (values whose configuration alters the behavior of the model). Traditionally, we use autocorrelation function (ACF) and PACF to find the optimal hyperparameters, which is subjective. Listing 2-12 uses the itertools package to find the best hyperparameters using Akaike information criterion, which measures out-of-sample predictions errors. itertools is pre-installed in the Python environment.

Listing 2-12. ARIMA Hyperparameters Optimization

```
from statsmodels.tsa.arima_model import ARIMA
import itertools
p = d = q = range(0, 2)
pdq = list(itertools.product(p, d, q))
```

```
for param in pdq:
    mod = ARIMA(train, order=param)
    results = mod.fit()
    print('ARIMA{} AIC:{}'.format(param, results.aic))

ARIMA(0, 0, 0) AIC:17153.28608377512
ARIMA(0, 0, 1) AIC:14407.085213632363
ARIMA(0, 1, 0) AIC:3812.4806641861296
ARIMA(0, 1, 1) AIC:3812.306176824848
ARIMA(1, 0, 0) AIC:3823.4611095477635
ARIMA(1, 0, 1) AIC:3823.432441560404
ARIMA(1, 1, 0) AIC:3812.267920836725
ARIMA(1, 1, 1) AIC:3808.9980010413774
```

Develop the ARIMA Model

Listing 2-13 completes the ARIMA (1, 1, 1) model and constructs profile tables for model performance evaluation (see Table 2-3). We choose only ARIMA (1,1,1) since it has the lowest AIC score. AIC is a statistical test that determines the goodness of fit and model simplicity. Simply, it indicates the extent to which a model loses information. So, order = (1,1,1) by default has a more predictive power than the other orders with the range in Listing 2-12.

Listing 2-13. Finalize the ARIMA Model

```
arima_model = ARIMA(train, order=(1, 1, 1))
arima_fitted = arima_model.fit()
arima_fitted.summary()
```

Table 2-3. *ARIMA Model Results*

Dep. Variable:	D.Adj Close	No. Observations:	2084
Model:	ARIMA(1, 1, 1)	Log Likelihood	-1900.499
Method:	css-mle	S.D. of innovations	0.602
Date:	Thu, 01 Apr 2021	AIC	3808.998
Time:	02:54:35	BIC	3831.566
Sample:	1	HQIC	3817.267

	coef	std err	z	P>|z|	[0.025	0.975]
const	0.0156	0.013	1.199	0.231	-0.010	0.041
ar.L1.D.Adj Close	-0.8638	0.094	-9.188	0.000	-1.048	-0.680
ma.L1.D.Adj Close	0.8336	0.103	8.095	0.000	0.632	1.035

	Real	Imaginary	Modulus	Frequency
AR.1	-1.1576	+0.0000j	1.1576	0.5000
MA.1	-1.1996	+0.0000j	1.1996	0.5000

Forecast Using the ARIMA Model

After completing the ARIMA (1, 1, 1) model, the subsequent step normally requires recognizing the underlying pattern of the adjusted close price and forecasting future prices (see Figure 2-8). See Listing 2-14.

Listing 2-14. Forecast ARIMA Model

```
fc, se, conf = arima_fitted.forecast(501, alpha=0.05)
fc_series = pd.Series(fc, index=test.index)
lower_series = pd.Series(conf[:, 0], index=test.index)
```

```
upper_series = pd.Series(conf[:, 1], index=test.index)
plt.plot(train, label="Training",color="navy")
plt.plot(test, label="Actual",color="orange")
plt.plot(fc_series, label="Forecast",color="red")
plt.fill_between(lower_series.index,
                 lower_series,
                 upper_series,
                 color='gray')
plt.legend(loc='upper left')
plt.xticks(rotation=45)
plt.xlabel("Date")
plt.ylabel("Adj Close")
plt.show()
```

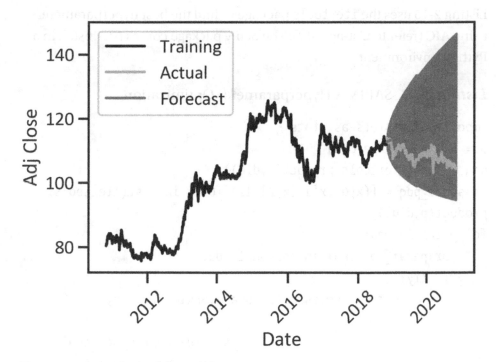

Figure 2-8. *ARIMA (1, 1, 1) forecast*

Figure 2-8 shows that the model makes errors when forecasting. The gray area (fill between) represents the confidence interval.

The SARIMA Model

Although the ARIMA model is a powerful model for analyzing patterns in univariate time-series data, it commits errors when handling seasonal data. Adding seasonal order to the model enhances its performance. SARIMA extends the ARIMA model. It considers the seasonal component when modeling time-series data.

SARIMA Hyperparameter Optimization

Listing 2-15 uses the `itertools` package to find the best hyperparameters using AIC (refer to Listing 2-12). The SciPy package comes pre-install in a Python environment.

Listing 2-15. SARIMA Hyperparameter Optimization

```
import scipy.stats as stats
p = d = q = range(0,2)
pdq = list(itertools.product(p,d,q))
seasonal_pdq = [(x[0],x[1],x[2],12) for x in list(itertools.
product(p,d,q))]
for param in pdq:
    for param_seasonal in seasonal_pdq:
        try:
            model = sm.tsa.statespace.SARIMAX(train,
                            order=param,
                            seasonal_order=param_seasonal,
                            enforce_stationarity=False,
                            enforce_intervibility=False)
```

```
        results = model.fit()
        print("SARIMAX {} x {} 12 - AIC: {}".
        format(param,param_seasonal,results.aic))
    except:
        continue
SARIMAX (1, 1, 1) x (0, 1, 1, 12) 12 - AIC: 3822.8419684760634
SARIMAX (1, 1, 1) x (1, 0, 0, 12) 12 - AIC: 3795.147211266854
SARIMAX (1, 1, 1) x (1, 0, 1, 12) 12 - AIC: 3795.0731989219726
SARIMAX (1, 1, 1) x (1, 1, 0, 12) 12 - AIC: 4580.905706671067
SARIMAX (1, 1, 1) x (1, 1, 1, 12) 12 - AIC: 3824.843959188799
```

Note that we showed only the last five outputs. The previous code estimates the AIC from SARIMAX $(0, 0, 0) \times (0, 0, 0, 12)$ 12 up until SARIMAX $(1, 1, 1) \times (1, 1, 1, 12)$ 12. We found that the SARIMAX $(1, 1, 1) \times (1, 1, 1, 12)$ 12 has the lowest AIC score. Listing 2-16 finalizes the SARIMA model with order = $(1,1,1)$.

Develop a SARIMA Model

Listing 2-16 completes the SARIMA model without enforcing stationary and invertibility and constructs a table with information about the model's performance (see Table 2-4).

Listing 2-16. Finalize the SARIMA Model

```
import pmdarima as pm
sarimax_model = pm.auto_arima(train, start_p=1, start_q=1,
start_P=1, start_Q=1,
                    max_p=5, max_q=5, max_P=5, max_Q=5,
                    seasonal=True,
```

```
                    stepwise=True, suppress_warnings=True,
                    D=10, max_D=10,
                    error_action='ignore')
sarimax_model.summary()
```

Table 2-4 shows that p-values of ar.L1, ma.L1, and sigma greater than 0.05. We can confirm that the series is stationary.

Table 2-4. *SARIMA Profile*

Dep. Variable:	Y	No. Observations:	2085
Model:	SARIMAX(1, 1, 1)	Log Likelihood	-1901.217
Date:	Sat, 14 Nov 2020	AIC	3808.435
Time:	02:01:47	BIC	3825.361
Sample:	0	HQIC	3814.637
	- 2085		
Covariance Type:	Opg		

	coef	std err	z	P>|z|	[0.025	0.975]
ar.L1	-0.8645	0.083	-10.411	0.000	-1.027	-0.702
ma.L1	0.8344	0.090	9.258	0.000	0.658	1.011
sigma2	0.3630	0.007	52.339	0.000	0.349	0.377

Ljung-Box (Q):	58.77	Jarque-Bera (JB):	955.16
Prob(Q):	0.03	Prob(JB):	0.00
Heteroskedasticity (H):	1.35	Skew:	0.04
Prob(H) (two-sided):	0.00	Kurtosis:	6.32

Forecast Using the ARIMA Model

Listing 2-17 constructs a plot that shows previously adjusted close prices of the USD/JPY pair and those predicted by the SARIMA (1, 1, 12) model (see Figure 2-9).

Listing 2-17. SARIMA Model Forecast

```
n_periods = 24
fitted, confint = sarimax_model.predict(n_periods=n_periods,
return_conf_int=True)
index_of_fc = pd.date_range(train.index[-1], periods =
n_periods, freq='MS')
fitted_series = pd.Series(fitted, index=index_of_fc)
lower_series = pd.Series(confint[:, 0], index=index_of_fc)
upper_series = pd.Series(confint[:, 1], index=index_of_fc)
plt.plot(train, label="Training",color="navy")
plt.plot(test, label="Actual",color="orange")
plt.plot(fitted_series, label="Forecast",color="red")
plt.fill_between(lower_series.index,
                 lower_series,
                 upper_series,
                 color='gray')
plt.legend(loc='upper left')
plt.xticks(rotation=45)
plt.xlabel("Date")
plt.ylabel("Adj Close")
plt.show()
```

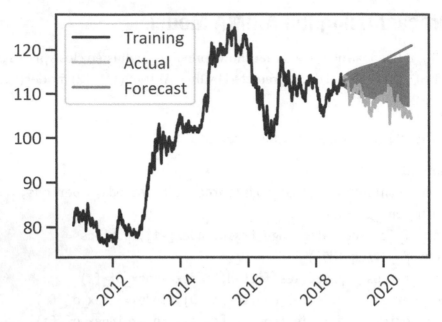

Figure 2-9. *SARIMA (1, 1, 1, 12) forecast*

Figure 2-9 shows a forecast with a narrow range. Both models do not best explain the time-series data; they commit marginal errors when forecasting future prices. In the next section, we will overcome this problem by using the additive model present on the Prophet package.

The Additive Model

Besides trends and seasonality, there are other factors that influence price changes. For instance, during public holidays, trading activities differ from normal trading days. Both the ARIMA and SARIMA models do not take into consideration the effects of public holidays. The additive model addressed this challenge. It considers daily, weekly, and yearly seasonality and nonlinear trends. It assumes that trends and cycles are one term and seamlessly adds the effects of official public holidays and an error. The formula is written mathematically as in Equation 2-1.

$$y = g(t) + s(t) + h(t) + \varepsilon_i \qquad \text{(Equation 2-1)}$$

Here, $g(t)$ represents the linear or logistic growth curve for modeling changes that are not periodic, $s(t)$ represents the periodic changes (daily, weekly, yearly seasonality), $h(t)$ represents the effects of holidays, and $+ \varepsilon i$ represents the error term that considers unusual changes.

Listing 2-18 repurposes the data (see Table 2-5). Please note that in Listing 2-2 we deleted columns of the low price, high price, open price, and close price. We are interested in forecasting the adjusted close price.

Listing 2-18. Data Preprocessing

```
df = df.reset_index()
df["ds"] = df["Date"]
df["y"] = df["Adj Close"]
df.set_index("Date")
```

Table 2-5. Dataset

Date	Adj Close	ds	y
2010-11-01	80.405998	2010-11-01	80.405998
2010-11-02	80.558998	2010-11-02	80.558998
2010-11-03	80.667999	2010-11-03	80.667999
2010-11-04	81.050003	2010-11-04	81.050003
2010-11-05	80.776001	2010-11-05	80.776001
...	...	...	...

(continued)

Table 2-5. *(continued)*

Date	Adj Close	ds	y
2020-10-27	104.832001	2020-10-27	104.832001
2020-10-28	104.544998	2020-10-28	104.544998
2020-10-29	104.315002	2020-10-29	104.315002
2020-10-30	104.554001	2020-10-30	104.554001
2020-11-02	104.580002	2020-11-02	104.580002

Listing 2-19 specifies the official public holidays whose effects will be added in the model (see Listing 2-20).

Listing 2-19. Specify Holidays

```
holidays = pd.DataFrame({
  'holiday': 'playoff',
  'ds': pd.to_datetime(["2020-12-25", "2020-12-24", "2020-12-23",
  "2019-12-25", "2021-01-01", "2021-01-20"]),
    "lower_window": 0,
    "upper_window": 1,
})
```

Listing 2-20 completes the additive model with a confidence interval of 95 percent; it considers yearly seasonality, weekly seasonality, daily seasonality, and official public holidays.

Listing 2-20. Develop Prophet Model

```
from fbprophet import Prophet

m = Prophet(holidays=holidays,
            interval_width=0.95,
            yearly_seasonality=True,
```

```
            weekly_seasonality=True,
            daily_seasonality=True,
            changepoint_prior_scale=0.095)
m.add_country_holidays(country_name='US')
m.fit(df)
```

Forecast

Listing 2-21 forecasts the future adjusted close price and shows patterns in the time-series data and in the prices that the additive model forecast (see Figure 2-10).

Listing 2-21. Forecast

```
future = m.make_future_dataframe(periods=365)
forecast = m.predict(future)
m.plot(forecast)
plt.xlabel("Date")
plt.ylabel("Adj Close")
plt.xticks(rotation=45)
plt.show()
```

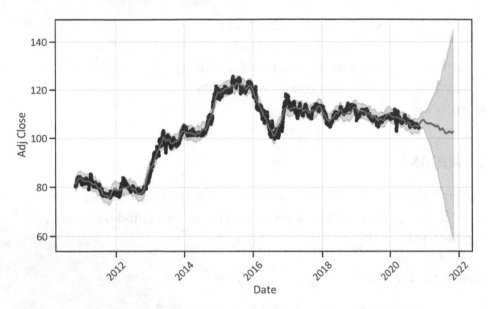

Figure 2-10. *Forecast*

Figure 2-10 disagrees with both the ARIMA model and SARIMA (both models predicted an upward trend). However, the additive model actually agrees with the data.

Seasonal Decomposition

Listing 2-22 applies the `plot_decompose()` method to decompose the series into seasonality, trend, and irregular components (see Figure 2-11). Decomposition involves breaking down a time series into components to understand the repeating patterns in the series. It helps determine parameters of univariate time-series analysis; we can identify whether there is a trend and seasonality in the series.

Listing 2-22. Seasonal Components

```
m.plot_components(forecast)
plt.show()
```

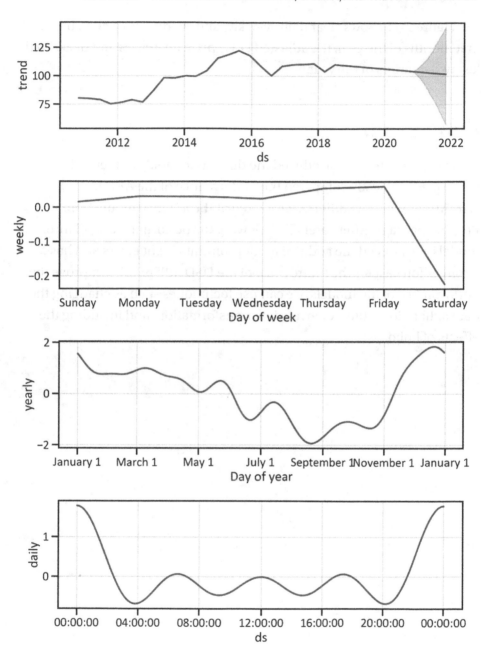

Figure 2-11. *Seasonal components*

Figure 2-11 shows clear daily, weekly, and yearly seasonality. In the first months of the year, the adjusted close price decreases and peaks in the last two quarters of the year.

Conclusion

This chapter carefully introduced the time-series analysis method. We developed and justly compared the performance of the ARIMA and SARIMA models. In addition, we looked at the additive model from the Prophet package. After carefully reviewing the performance of all three models, we noticed the additive model commits slight errors and has a stronger forecast for the future price of the USD/JPY pair. To improve the performance of the models, we can use techniques such as changing the data split ratio, outlier removal, data transformation, and including the effects of holidays.

CHAPTER 3

Univariate Time Series Using Recurrent Neural Nets

This chapter covers the basics of deep learning. First, it introduces the activation function, the loss function, and artificial neural network optimizers. Second, it discusses the sequence data problem and how a recurrent neural network (RNN) solves it. Third, the chapter presents a way of designing, developing, and testing the most popular RNN, which is the long short-term memory (LSTM) model. We use the Keras framework for rapid prototyping and building neural networks. To install `keras` in the conda environment, use `conda install -c conda-forge keras`. Ensure that you also install `tensorflow`. To install `tensorflow` in the conda environment, use `conda install -c conda-forge tensorflow`.

What Is Deep Learning?

Deep learning is a subset of machine learning that operates neural networks. A neural network is a network of interconnected groups of nodes that receive, transform, and transmit input values layer by layer until they reach the output layer. The activation function enables this process by operating a set of variables in each hidden layer.

© Tshepo Chris Nokeri 2021
T. C. Nokeri, *Implementing Machine Learning for Finance*,
https://doi.org/10.1007/978-1-4842-7110-0_3

Activation Function

The activation function adds nonlinearity to an artificial neural network and enables back propagation (a completed pass in reverse). There are three main activation functions.

> *The sigmoid activation function*: Fits an S-shaped curve to the data and triggers output values between 0 and 1.

> *The tangent hyperbolic (tanh) activation function*: Fits a tanh curve to the data and triggers outputs between -1 and 1.

> *The rectified linear unit (ReLu) activation function*: Retrieves values that are unconstrained by a specific range and addresses the vanishing gradient problem (a condition in which the gradient increases as we add more training data into the model, which results in slow training).

Loss Function

A loss function assesses the difference between the actual values and those predicted by an artificial neural network. Key loss functions include the mean squared error (MSE), mean absolute error, and mean squared logarithmic error. In this chapter, we use the MSE (the variability explained by the model about the data after considering a regression relationship).

Optimize an Artificial Neural Network

There are several methods for model optimization. The most common optimizer is the adaptive movement estimation (Adam), and it works better in minimizing the cost function during training.

The Sequential Data Problem

Sequential data encompasses order data points with some dependency. The time series is an excellent example of sequential data. In time-series data, each data point represents an observation at a certain period. Traditional neural networks like the feed-forward network encounter challenges when modeling sequential data because they cannot remember the preceding output values. This means that feed-forward networks merely produce output values without considering any dependencies in the data (a recurrent net combats this problem).

The RNN Model

A recurrent neural network model is applicable in sequential modeling. We call it *recurrent* because the model performs repetitive tasks for each data point in a series, whereby the output value depends on the preceding values. The decision outcomes of the recurrent network at time step $_{t-1}$ influence the decision outcomes at the last time step $_t$. It maintains a state to reference historical analysis at a given point. The state manages the information stored in preceding estimates and recurs back into a network with unique input values. Figure 3-1 depicts a recurrent with a single hidden layer.

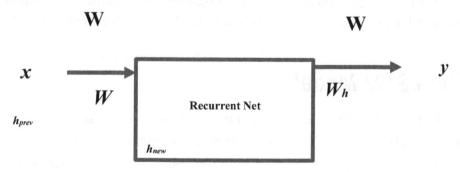

Figure 3-1. *RNN model*

The formula is expressed as shown in Equation 3-1.

$$h_{new} = tanh\left(W_h \times h_{prv} \times W_x \times x\right) \qquad \text{(Equation 3-1)}$$

Here, W_x represents the weight matrix between the input and the hidden unit, W_y represents ..., W_h represents weights multiplied by preceding states, x represents input data that the hidden layer receives, h_{prv} represents the preceding state the hidden layer receives, and h_{new} represents the new state that the hidden layer estimates. We apply the RNN model to speech recognition, image captioning, and sentiment analysis. Although the model combats problems that most models find difficult, it has its own drawbacks.

The Recurrent Neural Network Problem

The full RNN model must reasonably maintain a cell state. When dealing with big data, the network becomes computationally expensive; it is sensitive to alterations in parameters. Furthermore, it is prone to the vanishing gradient problem and the exploding gradient problem. This problem commonly occurs with traditional models, whereby at the initial phase of the training process, the model has a small gradient, but as we increase the training data, the gradient increases, resulting in a slower training process (this phenomenon is recognized as the vanishing gradient problem). The LSTM model combats the vanishing gradient problem.

The LSTM Model

The LSTM model can model long sequential data. It holds a strong gradient across several timestamps. There are two keys to the LSTM, namely, the cell state that moves information along without changes, and gates that control the information that flows alongside. See Figure 3-2.

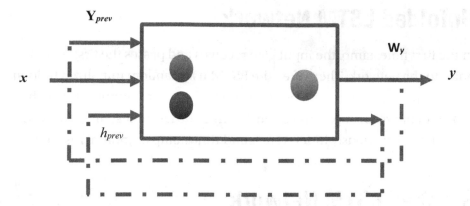

Figure 3-2. *LSTM*

Gates

The LSTM model encompasses three gates, namely, the input gate, the forget gate, and the output gate. The input gate determines additional information to be written in a memory cell. An input gate encompasses two layers. We recognize the first layer as the sigmoid layer. It determines the values that must be updated. The second layer is considered the tanh layer; it develops a vector of new values to include in a state. The forget gate determines the extent to which the model must forget and delete historical information no longer useful. Last, the output controls reading access from the memory cells. This is done by using estimates of the compressed function that is represented as (0,1), where 0 shows that the read access is denied and 1 indicates we grant read access. The previous gates are the operations of the LSTM that perform a function on a linear combination of inputs of the network, previously hidden state, and preceding output. The LSTM model uses gates to determine which data it should remember or forget.

Unfolded LSTM Network

In the first timestamp, the input gate receives and passes the first data point to the network. Thereafter, the LSTM uses random initialized hidden states to produce a new hidden state and transmits its output value to the subsequent timestamp. This continues until the last timestamp. The unit contains the previously hidden state and timestamp outputs across time.

Stacked LSTM Network

We recognize an LSTM network with more than one hidden layer as a stacked LSTM network. In a stacked LSTM network, it uses the output of a preceding hidden layer as the input of the subsequent layer. This process continues until the last layer. This allows for greater model complexity. For instance, subsequent layers are a more complex future representation of output compared to the preceding layers. Consequently, stacking an LSTM network may cause optimal model performance. In training, an LSTM does the following:

- Determines which data to add as inputs, including weights

- Estimates the new state based on the present and preceding internal state

- Learns corresponding weights and biases

- Determines how the state must be transmitted as output

Develop an LSTM Model Using Keras

Listing 3-1 applies the get_data_yahoo( ) method to extract Amazon stock prices[1] (see Table 3-1).

Listing 3-1. Scraped Data

```
from pandas_datareader import data
start_date = '2010-11-01'
end_date = '2020-11-01'
ticker = 'AMZN'
df = data.get_data_yahoo(ticker, start_date, end_date)
df.head()
```

---

[1]https://finance.yahoo.com/quote/AMZN

Table 3-1. Dataset

Date	High	Low	Open	Close	Volume	Adj Close
2010-11-01	164.580002	161.520004	164.449997	162.580002	5239900	162.580002
2010-11-02	165.940002	163.360001	163.750000	164.610001	4260000	164.610001
2010-11-03	168.610001	162.289993	165.399994	168.470001	6112100	168.470001
2010-11-04	172.529999	168.399994	169.860001	168.929993	7395900	168.929993
2010-11-05	171.649994	168.589996	169.350006	170.770004	5212200	170.770004

We are interested in studying the adjusted close price. Listing 3-2 creates a new dataframe that comprises only the adjusted closing price.

Listing 3-2. Create New Dataframe

```
df_close = pd.DataFrame(df["Adj Close"])
```

Listing 3-3 returns descriptive statistics of the adjusted closing price (see Table 3-2).

Listing 3-3. Descriptive Statistics

```
df_close.describe()
```

Table 3-2. *Descriptive Statistics*

	Adj Close
count	2518.000000
mean	881.182176
std	774.472276
min	157.779999
25%	269.540001
50%	551.135010
75%	1544.927551
max	3531.449951

Table 3-2 highlights that the mean value of the adjusted close is 881.18, and the standard deviation is 744.447. Listing 3-4 defines a functional argument to create attributes and find a list of instances, representing time without delays. First, we define the start date and end date. Thereafter, we set datetime as the index and create a copy of the dataframe so that we can create a list of previous instances from the specified start date and end date. Lastly, we create new columns with the frequent attributes and previous instances and merge the columns.

Listing 3-4. Create Regressor Attribute Function

```
def create_regressor_attributes(df, attribute, list_of_prev_t_
instants):
    list_of_prev_t_instants.sort()
    start = list_of_prev_t_instants[-1]
    end = len(df)
    df['datetime'] = df.index
    df.reset_index(drop=True)
    df_copy = df[start:end]
    df_copy.reset_index(inplace=True, drop=True)
    for attribute in attribute :
            foobar = pd.DataFrame()
            for prev_t in list_of_prev_t_instants :
                new_col = pd.DataFrame(df[attribute].
                iloc[(start - prev_t) : (end - prev_t)])
                new_col.reset_index(drop=True, inplace=True)
                new_col.rename(columns={attribute :
                '{}_(t-{})'.format(attribute, prev_t)},
                inplace=True)
                foobar = pd.concat([foobar, new_col],
                sort=False, axis=1)
            df_copy = pd.concat([df_copy, foobar], sort=False,
            axis=1)
    df_copy.set_index(['datetime'], drop=True, inplace=True)
    return df_copy
```

Listing 3-5 compiles a list of previous time instances.

Listing 3-5. List All Attributes

```
list_of_attributes = ['Adj Close']

list_of_prev_t_instants = []
for i in range(1,16):
    list_of_prev_t_instants.append(i)

list_of_prev_t_instants
[1, 2, 3, 4, 5, 6, 7, 8, 9, 10, 11, 12, 13, 14, 15]
```

The output shows a list of 15 instances. Listing 3-6 passes the list of attributes and list of previous t instances into a Pandas dataframe. Thereafter, it merges them with the adjusted closing price.

Listing 3-6. Create New Dataframe

```
df_new = create_regressor_attributes(df_close, list_of_
attributes, list_of_prev_t_instants)
```

Listing 3-7 imports key dependencies.

Listing 3-7. Import Important Libraries

```
from tensorflow.keras.layers import Input, Dense, Dropout
from tensorflow.keras.optimizers import SGD
from tensorflow.keras.models import Model
from tensorflow.keras.models import load_model
from tensorflow.keras.callbacks import ModelCheckpointfunction
```

Listing 3-8 creates the architecture of the neural network. We train the model with 15 variables using the linear activation in two dense layers and one output layer.

Listing 3-8. Design the Architecture

```
input_layer = Input(shape=(15), dtype='float32')
dense1 = Dense(60, activation='linear')(input_layer)
dense2 = Dense(60, activation='linear')(dense1)
dropout_layer = Dropout(0.2)(dense2)
output_layer = Dense(1, activation='linear')(dropout_layer)
```

Listing 3-9 trains and summarizes the model. To determine the extent to which the model makes correct predictions in the training process, we use the mean squared error, representing variability explained in the model after we consider a linear relationship. To improve the performance of the model, we use the Adam optimizer, which is an optimizer that considers the force that keeps the gradient moving and lowers the rate at which the model learns the data.

Listing 3-9. Network Structure

```
model = Model(inputs=input_layer, outputs=output_layer)
model.compile(loss='mean_squared_error', optimizer='adam')
model.summary()
Model: "model"
```

Layer (type)	Output Shape	Param #
input_1 (InputLayer)	[(None, 15)]	0
dense (Dense)	(None, 60)	960
dense_1 (Dense)	(None, 60)	3660
dropout (Dropout)	(None, 60)	0

| dense_2 (Dense) | (None, 1) | 61 |

==

Total params: 4,681
Trainable params: 4,681
Non-trainable params: 0

The neural networks encompass two hidden layers and a dropout layer (applying the probability of setting each input of the layer to 0). Listing 3-10 splits the data into training, test data, and validation data.

Listing 3-10. Split Data into Training, Test, and Validation Data

```
test_set_size = 0.05
valid_set_size= 0.05
df_copy = df_new.reset_index(drop=True)
df_test = df_copy.iloc[ int(np.floor(len(df_copy)*(1-test_set_
size))) : ]
df_train_plus_valid = df_copy.iloc[ : int(np.floor(len(df_
copy)*(1-test_set_size))) ]
df_train = df_train_plus_valid.iloc[ : int(np.floor(len(df_
train_plus_valid)*(1-valid_set_size))) ]
df_valid = df_train_plus_valid.iloc[ int(np.floor(len(df_train_
plus_valid)*(1-valid_set_size))) : ]
X_train, y_train = df_train.iloc[:, 1:], df_train.iloc[:, 0]
X_valid, y_valid = df_valid.iloc[:, 1:], df_valid.iloc[:, 0]
X_test, y_test = df_test.iloc[:, 1:], df_test.iloc[:, 0]
print('Shape of training inputs, training target:', X_train.
shape, y_train.shape)
print('Shape of validation inputs, validation target:',
X_valid.shape, y_valid.shape)
print('Shape of test inputs, test target:', X_test.shape,
y_test.shape)
```

Here, we have this information:

- *Shape of training inputs, training target*: (2258, 15) (2258,)

- *Shape of validation inputs, validation target*: (119, 15) (119,)

- *Shape of test inputs, test target*: (126, 15) (126,)

Listing 3-11 applies the MinMaxScaler() method to scale numeric values between 0 and 1.

Listing 3-11. Normalize Data

```
from sklearn.preprocessing import MinMaxScaler
Target_scaler = MinMaxScaler(feature_range=(0.01, 0.99))
Feature_scaler = MinMaxScaler(feature_range=(0.01, 0.99))
X_train_scaled = Feature_scaler.fit_transform(np.array(X_
train))
X_valid_scaled = Feature_scaler.fit_transform(np.array(X_
valid))
X_test_scaled = Feature_scaler.fit_transform(np.array(X_test))
y_train_scaled = Target_scaler.fit_transform(np.array(y_train).
reshape(-1,1))
y_valid_scaled = Target_scaler.fit_transform(np.array(y_valid).
reshape(-1,1))
y_test_scaled = Target_scaler.fit_transform(np.array(y_test).
reshape(-1,1))Develop the LTSM Model
```

Listing 3-12 trains the LTSM model across 30 epochs (complete forward and backward passes) in 5 batches (sets of sample trains at one time). A pass represents the complete training iteration, from receiving a set of input values to firing them with different weights and biases across the network until an output value is produced.

Listing 3-12. Train the Recurrent Network

```
history = model.fit(x=X_train_scaled, y=y_train_scaled, batch_
size=5, epochs=30, verbose=1, validation_data=(X_valid_scaled,
y_valid_scaled), shuffle=True)
```

Forecasting Using the LTSM

Listing 3-13 applies the predict() method to forecast future instances of the series and perform an inverse transformation (produces exponential random variables). See Table 3-3.

Listing 3-13. LTSM Forecast

```
y_pred = model.predict(X_test_scaled)
y_test_rescaled =  Target_scaler.inverse_transform(y_test_
scaled)
y_pred_rescaled = Target_scaler.inverse_transform(y_pred)
y_actual = pd.DataFrame(y_test_rescaled, columns=['Actual Close
Price'])
y_hat = pd.DataFrame(y_pred_rescaled, columns=['Predicted Close
Price'])
pd.DataFrame(y_pred_rescaled, columns = ["Forecast"]).head()
```

Table 3-3. Forecast

	Forecast
0	2370.852783
1	2355.488281
2	2338.975586
3	2380.348633
4	2397.861084

Listing 3-14 shows the actual value of the adjusted close prices= and the values that the LSTM model forecasts (see Figure 3-3).

Listing 3-14. Forecast

```
plt.plot(y_actual,color='red')
plt.plot(y_hat, linestyle='dashed', color='navy')
plt.legend(['Actual','Predicted'], loc='best')
plt.ylabel('Adj Close')
plt.xlabel('Test Set Day no.')
plt.xticks(rotation=45)
plt.yticks()
plt.show()
```

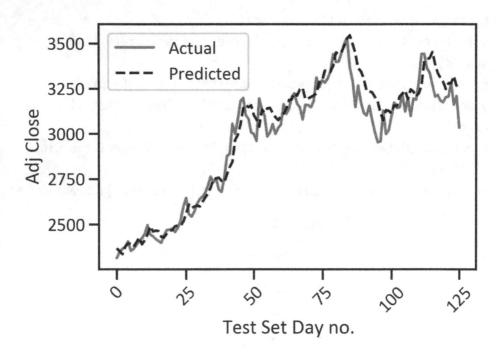

Figure 3-3. *Forecast*

Figure 3-3 shows there is a difference between the actual values of the adjusted closing price and the values that the LSTM model forecasts. Moreover, the difference is not significant enough to impact conclusions.

Model Evaluation

Table 3-4 highlights key metrics we used to evaluate a classifier.

Table 3-4. *Key Evaluation Metrics*

Metric	Description
Mean absolute error (MAE)	The average degree of error in estimates without considering the direction
Mean squared error	The variability explained by the model about the data after considering a regression relationship
Root mean squared error	The variability explained without considering a regression relationship
R-squared	The variability explained by the model about the data

Listing 3-15 returns a table with key regression evaluation metrics (see Table 3-5).

Listing 3-15. Develop a Model Evaluation Matrix

```
from sklearn import metrics
MAE = metrics.mean_absolute_error(y_test_rescaled,y_pred_
rescaled)
MSE = metrics.mean_squared_error(y_test_rescaled,y_pred_
rescaled)
```

```python
RMSE = np.sqrt(MSE)
R2 = metrics.r2_score(y_test_rescaled,y_pred_rescaled)
EV = metrics.explained_variance_score(y_test_rescaled,y_pred_
rescaled)
MGD = metrics.mean_gamma_deviance(y_test_rescaled,y_pred_
rescaled)
MPD = metrics.mean_poisson_deviance(y_test_rescaled,y_pred_
rescaled)
lmmodelevaluation = [[MAE,MSE,RMSE,R2,EV,MGD,MPD]]
lmmodelevaluationdata = pd.DataFrame(lmmodelevaluation,
                                 index = ["Values"],
                                 columns = ["MAE",
                                            "MSE",
                                            "RMSE",
                                            "R2",
                                            "Explained
                                            variance
                                            score",
                                            "Mean gamma
                                            deviance",
                                            "Mean Poisson
                                            deviance"]).
                                            transpose()

lmmodelevaluationdata
```

Table 3-5. *Model Performance*

	Values
MAE	60.672022
MSE	6199.559394
RMSE	78.737281
R2	0.944119
Explained variance score	0.944151
Mean gamma deviance	0.000657
Mean Poisson deviance	2.010341

Table 3-5 highlights that the LSTM model explains 94.44 percent of the variability in the data. The average magnitude of the error without considering the direction of the relation is 60.67. Another way of evaluating the performance of an LSTM model involves assessing changes in the loss across different epochs. Loss measures the difference between the actual values and the values that the models predict. Figure 3-4 depicts how the LSTM model learns how to differentiate between the actual values and the predicted values. See Listing 3-16.

Listing 3-16. Training and Validation Loss Across Epochs

```
plt.plot(history.history["loss"],color="red",label="Training
Loss")
plt.plot(history.history["val_loss"],color="navy",label="Cross-
Validation Loss")
plt.xlabel("Epochs")
plt.ylabel("Loss")
plt.legend(loc="best")
plt.show()
```

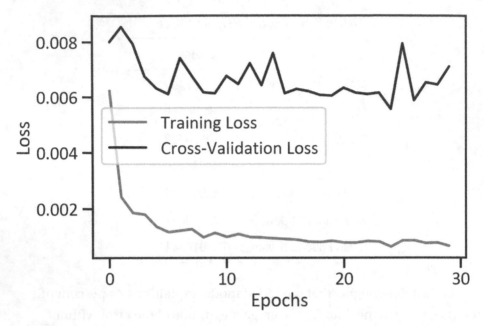

Figure 3-4. *Training and validation loss across epochs*

Figure 3-4 shows that at the first epoch the training loss drops sharply. Cross-validation loss is stable across time; it is also above training loss across different epochs. The LSTM model shows the characteristics of a well-behaved model.

Conclusion

This chapter introduced deep learning. It covered an artificial neural network model that we used for time-series data recognized as the LSTM model. It also showed a way to create variables for modeling and showed the structure of the neural network. Thereafter, it covered techniques for modeling and testing the network.

After reviewing the network, we found that the model best explains the variability in the data. We can use this model to forecast future instances of the adjusted closing price. We can improve the performance of the network by reducing layers, introducing a penalty term, etc. The next chapter covers ways of identifying hidden patterns in sequential data using hidden Markov models.

CHAPTER 4

Discover Market Regimes

This chapter presents a widespread generative probabilistic model called the *hidden Markov model* (HMM). It covers effective techniques for identifying hidden patterns in time-series data using HMM and generating a sequence of observations. After reading this chapter, you will be able to design and develop a hidden Markov model with a Gaussian process to discover market regimes. To install `hmmlearn` in the Python environment, use `pip install hmmlearn`, and in the conda environment, use `conda install -c conda-forge hmmlearn`.

HMM

In the preceding chapter, we used the augmented Dickey-Fuller test (a unit root test) to inspect and determine whether a series was stationary. A series is stationary when the mean value does not vary across time. In that chapter, we found the series to be not stationary. In this chapter, we use the hidden Markov model. We can use this model for a stationary or nonstationary series. It is a generative, unsupervised learning model; it does not require labels, and we resort to it for unstructured data. We

© Tshepo Chris Nokeri 2021
T. C. Nokeri, *Implementing Machine Learning for Finance*,
https://doi.org/10.1007/978-1-4842-7110-0_4

call it *generative* because it generates data after uncovering patterns in sequential data. You will learn more about HMM as you progress through this chapter.

HMM Application in Finance

We frequently use HMM for speech analysis, weather forecasting, etc. In this chapter, we will apply HMM to uncover hidden patterns in the stock prices of the S&P 500 index. We will exploit this model to detect market regimes (periods in which there is low or high volatility in the market) and future regimes.

Develop a GaussianHMM

There are two principal reasons we will apply the Gaussian mixture model in this chapter. To begin with, we are interested in the likelihood of a series given a class representing two conditions: a condition in which the market is rising and expected to reasonably rise (called a *bullish* trend) and a condition in which the market is declining and is expected to decline (called a *bear* trend). Second, we assume that the observations are Gaussian with known variance and mean. This is referred to as a *Gaussian mixture* because of the shared distributions that are constrained to Gaussian. More generally, we will unanimously predict future classes of the trend based on observed classes. In the training process, the model learns a single Gaussian state output distribution. Thereafter, a distribution with the highest variance is fragmented, and the process iterates until a junction. Listing 4-1 shows the price data of the Standard & Poor (S&P)

500 index,[1] which is a stock market index that serves as a benchmark for 500 global companies listed on the US stock exchange. Table 4-1 shows the price data from November 1, 2010, to November 2, 2020.

Listing 4-1. Scraped Data

```
from pandas_datareader import data
from datetime import datetime
ticker = "^GSPC"
start_date = datetime.date(2010, 11, 1)
end_date = datetime.date(2020, 11, 1)
df = data.DataReader(ticker, 'yahoo', start_date, end_date)
df.head()
```

[1]https://finance.yahoo.com/quote/%5EGSPC/

Table 4-1. Dataset

Date	High	Low	Open	Close	Volume	Adj Close
2010-11-01	1195.810059	1177.650024	1185.709961	1184.380005	4129180000	1184.380005
2010-11-02	1195.880005	1187.859985	1187.859985	1193.569946	3866200000	1193.569946
2010-11-03	1198.300049	1183.560059	1193.790039	1197.959961	4665480000	1197.959961
2010-11-04	1221.250000	1198.339966	1198.339966	1221.060059	5695470000	1221.060059
2010-11-05	1227.079956	1220.290039	1221.199951	1225.849976	5637460000	1225.849976

Listing 4-2 applies the describe() function to get basic statistical results about the S&P 500 index (see Table 4-2).

Listing 4-2. Descriptive Statistics

```
df.describe()
```

Table 4-2. Descriptive Statistics

	High	Low	Open	Close	Volume	Adj Close
count	2518.000000	2518.000000	2518.000000	2518.000000	2.518000e+03	2518.000000
mean	2140.962839	2118.590950	2130.285829	2130.652759	3.737823e+09	2130.652759
std	620.951490	614.925344	618.284893	617.991582	8.780746e+08	617.991582
min	1125.119995	1074.770020	1097.420044	1099.229980	1.025000e+09	1099.229980
25%	1597.827454	1583.157501	1592.527527	1593.429993	3.242805e+09	1593.429993
50%	2085.185059	2066.575073	2077.265015	2078.175049	3.593470e+09	2078.175049
75%	2689.744934	2656.970032	2677.815063	2673.570068	4.056060e+09	2673.570068
max	3588.110107	3535.229980	3564.739990	3580.840088	9.044690e+09	3580.840088

Table 4-2 gives us an idea of the central tendency and dispersion of the price and volume. It highlights that the mean value of the adjusted closing price is 2130.652759, and the standard deviation is 617.991582. In relation to the volume, the maximum volume is 9.044690e+09, and the minimum volume is 1.025000e+09. The average low price is 2118.590950, and the average high price is 2140.962839. After performing a descriptive analysis, we start the data preprocessing process. Listing 4-3 drops the irrelevant variables.

Listing 4-3. Initial Data Preprocessing

```
df.reset_index(inplace=True,drop=False)
df.drop(['Open','High','Low','Adj Close'],axis=1,inplace=True)
df['Date'] = df['Date'].apply(datetime.datetime.toordinal)
df = list(df.itertuples(index=False, name=None))
```

Listing 4-4 assigns arrays.

Listing 4-4. Final Data Preprocessing

```
dates = np.array([q[0] for q in df], dtype=int)
end_val = np.array([q[1] for q in df])
volume = np.array([q[2] for q in df])[1:]
```

Listing 4-5 shows "differenced" time-series data (a series without temporal dependence).

Listing 4-5. Time Series

```
from matplotlib.dates import YearLocators
diff = np.diff(end_val)
dates = dates[1:]
end_val = end_val[1:]
X = np.column_stack([diff, volume])
fig, ax = plt.subplots()
```

```
plt.gca().xaxis.set_major_locator(YearLocator())
plt.plot_date(dates,end_val,"-",color="navy")
plt.xticks(rotation=45)
plt.xlabel("Date")
plt.ylabel("Adj Close")
plt.show()
```

Figure 4-1 shows that after the acute crisis, the longest bullish market
in history started emerging, which was inevitably followed by a minor
correction at the beginning of 2020. Afterward, the price increased in
massive increments. The market hit 3,400 in October and rallied upward.

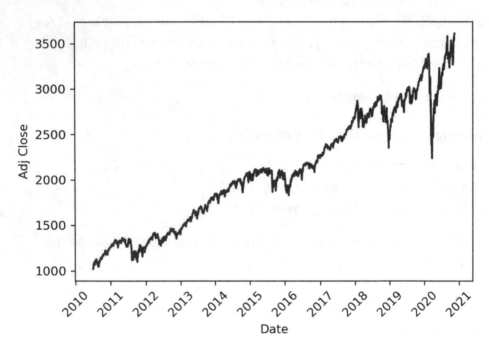

Figure 4-1. *Time series*

Gaussian Hidden Markov

The Gaussian hidden Markov model (GaussianHMM) assumes that the underlying structure of the probabilistic distributions is normal. We use GaussianHMM when dealing with a continuous variable that comes from a normal distribution. In the preceding chapter, we found that the time-series data has a serial correlation across different lags. GaussianHMM is more suitable for the problem at hand. HMM represents an unsupervised learning model for unraveling sequential problems. In unsupervised learning, we do not hide any data away from the model; we expose the model to all the data. It does not require us to split the data into training and test data. Now that we recognize the pattern of the time-series data, we can proceed and complete the model. Listing 4-6 applies the `fit()` method to complete a Gaussian hidden Markov model (with a spectral mixture kernel that produces Gaussian emissions). It specifies the number of components as 5, the number of iterations as 10, and the total as 0.0001.

Listing 4-6. Finalize the GaussianHMM

```
model = GaussianHMM(n_components=5, covariance_type="diag",
n_iter=1000)
model.fit(X)
```

Listing 4-7 predicts the sequence of internal hidden states. Thereafter, it tabulates the sequence and shows the first five predicted hidden states (see Table 4-3).

Listing 4-7. Hidden States

```
hidden_states = model.predict(X)
pd.DataFrame(hidden_states,columns=["hidden_state"])
```

Table 4-3. *Hidden States*

	hidden_state
0	1
1	1
2	1
3	1
4	1
...	...
2512	1
2513	1
2514	3
2515	1
2516	1

Table 4-3 highlights the parameters of the most appropriate sequence of hidden states. It does not provide sufficient information to conclude how GaussianHMM models the time-series data. Listing 4-8 plots the sequence of internal hidden states of S&P 500 time-series data that the GaussianHMM model produced (see Figure 4-2).

Listing 4-8. HMM Results

```
num_sample = 3000
sample, _ = model.sample(num_sample)
plt.plot(np.arange(num_sample), sample[:,0],color="navy")
plt.xlabel("Samples")
plt.ylabel("States")
plt.xticks(rotation=45)
plt.show()
```

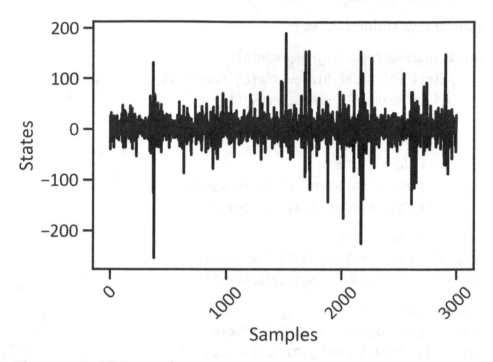

Figure 4-2. *HMM results*

Figure 4-2 shows the sample data. At most, there is stability in the states. After the 500th observation, there was a slight spike. The most significant spike after the sample was 1,200.

Mean and Variance

The mean value gives us a lot of information about the central tendency of data points, and the variance shows the dispersion of data points away from the mean value. We use both the mean value and the variance to summarize the optimal sequence of the hidden states that the GaussianHMM model produced. Listing 4-9 estimates the mean value and variance of each internal hidden state.

Listing 4-9. Hidden States

```
for i in range(model.n_components):
    print("{0} order hidden state".format(i))
    print("mean = ", model.means_[i])
    print("var = ", np.diag(model.covars_[i]))
    print()
0 order hidden state
mean =  [-8.29728342e-01  4.41641901e+09]
var =  [8.50560173e+02 4.47992619e+17]

1 order hidden state
mean =  [2.22719353e+00 3.22150428e+09]
var =  [1.33720438e+02 6.16375025e+16]

2 order hidden state
mean =  [2.04319405e+00 3.73786775e+09]
var =  [1.79332780e+02 9.03360887e+16]

3 order hidden state
mean =  [9.56404042e-01 2.50937758e+09]
var =  [8.46343104e+01 3.11987532e+17]

4 order hidden state
mean =  [-1.12364778e+01  6.07623360e+09]
var =  [8.06282637e+03 1.77082805e+18]
```

Listing 4-10 plots the sequence of each internal hidden state (see Figure 4-3).

Listing 4-10. Individual Sequence of Hidden States

```
fig, axs = plt.subplots(model.n_components, sharex=True,
sharey=True, figsize=(15,15))
colours = cm.rainbow(np.linspace(0, 1, model.n_components))
for i, (ax, colour) in enumerate(zip(axs, colours)):
```

```
    mask = hidden_states == i
    ax.plot_date(dates[mask], end_val[mask], ".", c=colour)
    ax.set_title("{0}th hidden state".format(i))
    ax.set_xlabel("Date")
    ax.set_ylabel("Adj Close")
    ax.xaxis.set_major_locator(YearLocator())
    ax.xaxis.set_minor_locator(MonthLocator())
    ax.grid(True)
plt.show()
```

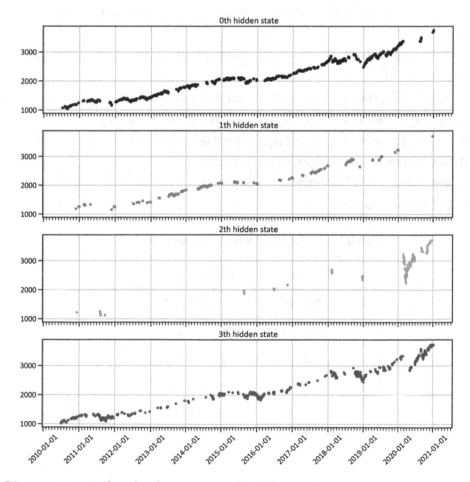

Figure 4-3. *Individual sequence of hidden states*

Figure 4-3 shows the volatility in the upward trend. Most of 2020 happened in the third hidden state.

Expected Returns and Volumes

Listing 4-11 tabulates the expected returns and volumes estimated by the GaussianHMM (see Table 4-4).

Listing 4-11. Expected Returns and Volumes

```
expected_returns_and_volumes = np.dot(model.transmat_, model.
means_)
returns_and_volume_columnwise = list(zip(*expected_returns_and_
volumes))
expected_returns = returns_and_volume_columnwise[0]
expected_volumes = returns_and_volume_columnwise[1]
params = pd.concat([pd.Series(expected_returns),
pd.Series(expected_volumes)], axis=1)
params.columns= ['Returns', 'Volume']
pd.DataFrame(params)
```

Table 4-4. *Expected Returns and Volume*

	Returns	Volume
0	1.789235	3.752679e+09
1	-0.433996	4.258633e+09
2	2.013629	3.288045e+09
3	-9.934745	5.868430e+09
4	1.139917	2.752306e+09

Listing 4-12 creates a dataframe for future returns and volumes, and it plots the actual values of the adjusted close price and those predicted by the GaussianHMM (see Figure 4-4).

Listing 4-12. Actual Price and Predicted Price

```
lastN = 7
start_date = datetime.date.today() - datetime.
timedelta(days=lastN*2)
dates = np.array([q[0] for q in df], dtype=int)
predicted_prices = []
predicted_dates = []
predicted_volumes = []
actual_volumes = []
for idx in range(lastN):
    state = hidden_states[-lastN+idx]
    current_price = df[-lastN+idx][1]
    volume = df[-lastN+idx][2]
    actual_volumes.append(volume)
    current_date = datetime.date.fromordinal(dates[-lastN+idx])
    predicted_date = current_date + datetime.timedelta(days=1)
    predicted_dates.append(predicted_date)
    predicted_prices.append(current_price + expected_
    returns[state])
    predicted_volumes.append(np.round(expected_volumes[state]))
fig, ax = plt.subplots()
plt.plot(predicted_dates,end_val[-lastN:],color="navy",label="A
ctual Price")
plt.plot(predicted_dates,predicted_prices,color="red",label="Pr
edicted Price")
plt.legend(loc="best")
plt.xticks(rotation=45)
```

```
plt.xlabel("Date")
plt.ylabel("Adj Close")
plt.show()
```

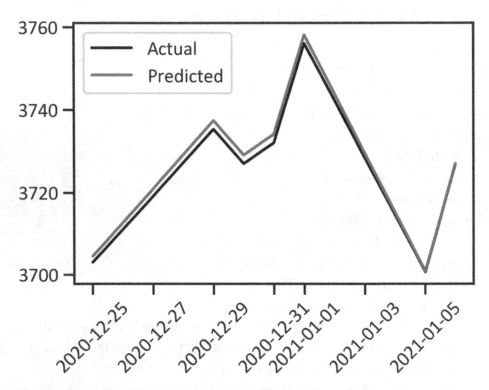

Figure 4-4. *Actual price and predicted price*

Figure 4-4 displays the notable characteristics of a well-behaved GaussianHMM model. The differences between the actual values of the adjusted closing price and predicted values are small. Listing 4-13 plots the actual volume and the predicted volume (see Figure 4-5).

Listing 4-13. Actual Volume and Predicted Volume

```
fig, ax = plt.subplots()
plt.plot(predicted_dates,actual_volumes,color="navy",label="Act
ual Volume")
plt.plot(predicted_dates,predicted_volumes,color="red",label="P
redicted Volume")
plt.legend(loc="best")
plt.xticks(rotation=45)
plt.xlabel("Date")
plt.ylabel("Volume")
plt.show()
```

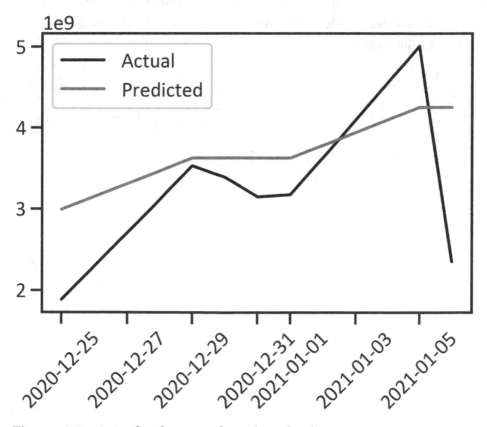

Figure 4-5. *Actual volume and predicted volume*

GaussianHMM does well in modeling the time-series data; however, there are minimal differences between the actual values of the volume and those predicted by the model.

Conclusions

This chapter presented the hidden Markov model. It artfully covered practical ways of designing and developing an HMM with a spectral kernel that produces Gaussian emissions to combat complex sequential problems. We rigorously applied the model to the time-series data of the S&P 500 index to estimate the best sequence of internal hidden states. After carefully reviewing the performance of the model, we noticed that the model is skillful in identifying hidden patterns in the price and volume of the S&P 500 index.

CHAPTER 5

Stock Clustering

A combination of stocks that an investor selects influences the investment portfolio's performance. Some assets are highly risky to invest in, but others are not. High-risk assets are those whose prices change drastically in short periods. To secure capital, investors may select groups of stocks, rather than investing in a single stock or class of stocks. Given that there are many stocks to choose from, investors ordinarily find it arduous to single out a group of stocks with optimal performance. To find a group of stocks with similarities, we use an unsupervised learning technique called *cluster analysis.* It involves grouping data points based on similar characteristics. The most popular cluster analysis model is the k-means model.

Investment Portfolio Diversification

Investors ought to be careful and not put all their eggs in one basket by directing all their investment funds to one area of interest. They may use portfolio diversification to secure capital. It involves allocating funds to assets across various industries, sectors, geographic locations, etc., and a conservative investor diversifies their portfolio by investing in less risky and more profitable stocks. They ensure that their strategy matches their risk appetite and invest in the best-performing stocks with minimum volatility.

© Tshepo Chris Nokeri 2021
T. C. Nokeri, *Implementing Machine Learning for Finance,*
https://doi.org/10.1007/978-1-4842-7110-0_5

Stock Market Volatility

We mentioned in the preceding chapters that prices in the foreign exchange market and the stock market are not constant. There is an element of randomness because they fluctuate across time. In terms of liquidity, the foreign exchange market is the most liquid market, followed by the stock market, and so on. Market makers drive price movements by the volume of large transactions and sales activities.

Investing in contracts for difference (CFD) and exchange-traded funds (EFTs) involves risk. The price of an asset may go in the opposite direction, resulting in capital loss. Investors ought to have robust risk management strategies. Ideally, they must select the best stocks based on certain criteria. The most common criterion is volatility, which is an estimate of the degree of drastic changes in prices in a short period. We basically assume that if liquidity increases in a market, then volatility increases, and vice versa. We measure volatility by estimating the standard deviation of logarithmic returns and beta coefficients. Besides large transactions and sales activities of multinational financial institutions, there are other factors that influence volatility such as social events, economic events, holidays, pandemics, labor unrest, natural disasters, war, etc.

K-Means Clustering

The k-means model partitions the data into k (clusters) with the nearest mean (centroids); it then finds the distance between subgroups to produce a cluster. It simultaneously shrinks the intracluster distances and improves the intercluster. We express the formula as shown in Equation 5-1.

$$Dis(x_1, x_2) = \sqrt{\sum_{i=0}^{n}(x_{1i} - y_{2i})^2} \qquad \text{(Equation 5-1)}$$

Dis(x_1,x_2) reflects the distance between the data points. Based on the formula in Equation 5-1, we are interested in discovering the square root of the sum squares of the deviation of independent data points (represented as coordinates (x_1, y_1) and (x_2, y_2)) away from the mean value (the center). It finds the initial *k* (the number of clusters) and estimates the distance between clusters, and then it distributes data points to the adjoining centroid. It splits the data points into *k* groups of similarities. The most popular method of estimating similarities in space is Euclidean distance (which connects angles and distances). This cluster model requires the numbers to specify on the set. Data partitioning relies on a number of clusters. This algorithm randomly initializes centroids.

K-Means in Practice

Listing 5-1 extracts the data from Yahoo Finance and applies the web scraping get_data_yahoo() method. Thereafter, it performs the data processing tasks required to have quality data.

Listing 5-1. Scraped Data

```
rom pandas_datareader import data
tickers = ['AMZN','AAPL','WBA',
           'NOC','BA','LMT',
           'MCD','INTC','NAV',
           'IBM','TXN','MA',
           'MSFT','GE','AXP',
           'PEP','KO','JNJ',
           'TM','HMC','MSBHY',
           'SNE','XOM','CVX',
           'VLO','F','BAC']
```

```
start_date = '2010-01-01'
end_date = '2020-11-01'
df = data.get_data_yahoo(tickers, start_date, end_date)[['Adj
Close']]
```

Listing 5-1 extracts stocks from companies such as Amazon, Apple, and Walgreens Boots Alliance, among other stocks. There are 27 stocks. Remember that you may include as many stocks as you want. Listing 5-2 estimates the return and volatility.

Listing 5-2. Estimate Returns and Volatility

```
returns = df.pct_change().mean() * (10*12)
std = df.pct_change().std() * np.sqrt((10*12))
ret_var = pd.concat([returns, std], axis = 1).dropna()
ret_var.columns = ["Returns","Standard Deviation"]
```

Listing 5-3 is an elbow curve (see Figure 5-1). We use it to determine the number of clusters to specify when we develop the k-means model.

Listing 5-3. Elbow Curve

```
X =  ret_var.values
sse = []
for k in range(1,15):
    kmeans = KMeans(n_clusters = k)
    kmeans.fit(X)
    sse.append(kmeans.inertia_)
plt.plot(range(1,15), sse)
plt.xlabel("Value of k")
plt.ylabel("Distortion")
plt.show()
```

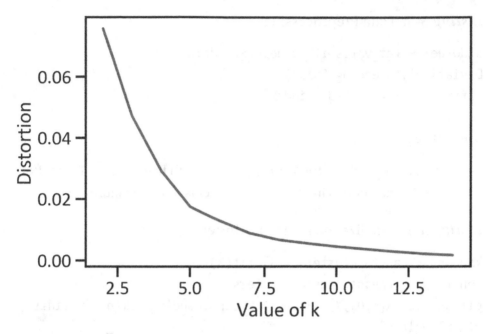

Figure 5-1. *Elbow curve*

The y-axis shows the compressed variance of the correlation matrix (eigenvalues), and the x-axis shows the number of factors. We use the figure to determine the required number of clusters for a cluster model by finding a point in a curve where a sharp decline begins. To clearly understand how this works, consider the y-axis as the severity of the correlation. We are interested in the borderline between severe correlation and nonsevere correlation. Figure 5-1 displays a smooth bend from 1 to 5. However, from 5, the curve bends abruptly. We use 5 as the cutoff point. Listing 5-4 sorts the standard deviation by descending order, removes all missing values, and creates a NumPy array of the Pandas dataframe.

Listing 5-4. Data Preprocessing

```
stdOrder = ret_var.sort_values('Standard
Deviation',ascending=False)
first_symbol = stdOrder.index[0]
ret_var.drop(first_symbol,inplace=True)
X = ret_var.values
```

Listing 5-5 completes the k-means model with five clusters. Thereafter, it depicts the data points in their respective clusters (see Figure 5-2).

Listing 5-5. Finalize the K-Means Model

```
kmeans =KMeans(n_clusters = 5).fit(X)
centroids = kmeans.cluster_centers_
plt.scatter(X[:,0],X[:,1], c = kmeans.labels_, cmap ="viridis")
plt.xlabel("y")
plt.scatter(centroids[:,0], centroids[:,1],color="red",mark
er="*")
plt.show()
```

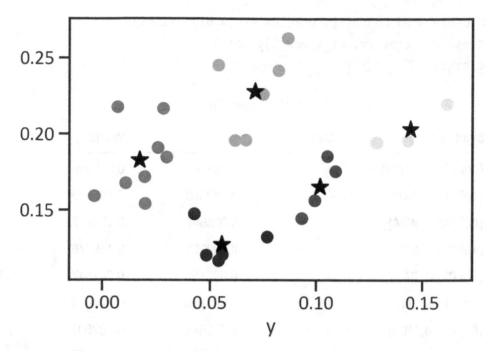

Figure 5-2. *K-means model*

The k-means model made intelligent guesstimates until it allocated data points to the most adjacent centroid and discovered the mean value of the centroids. Figure 5-2 shows that there are five apparent clusters in the data. Listing 5-6 tabulates each stock, with the cluster it belongs to, together with its returns and volatility (see Table 5-1).

Listing 5-6. Returns and Volatility per Cluster

```
stocks = pd.DataFrame(ret_var.index)
cluster_labels = pd.DataFrame(kmeans.labels_)
stockClusters = pd.concat([stocks, cluster_labels],axis = 1)
stockClusters.columns = ['Symbol','Cluster']
x_df = pd.DataFrame(X, columns = ["Returns", "Volatitity"])
```

```
closerv = pd.concat([stockClusters,x_df],axis=1)
closerv = closerv.set_index("Symbol")
closerv
```

Table 5-1. *Returns and Volatility per Cluster*

Symbol	Cluster	Returns	Volatility
(Adj Close, AMZN)	4	0.161408	0.219298
(Adj Close, AAPL)	4	0.142889	0.195379
(Adj Close, WBA)	2	0.025647	0.190772
(Adj Close, NOC)	1	0.099201	0.156376
(Adj Close, BA)	3	0.082153	0.241547
(Adj Close, LMT)	1	0.093110	0.144674
(Adj Close, MCD)	0	0.076895	0.132607
(Adj Close, INTC)	3	0.066716	0.195823
(Adj Close, IBM)	2	0.019773	0.154294
(Adj Close, TXN)	1	0.105104	0.185222
(Adj Close, MA)	4	0.128066	0.194152
(Adj Close, MSFT)	1	0.108985	0.175395
(Adj Close, GE)	2	0.006920	0.217385
(Adj Close, AXP)	3	0.061691	0.195526
(Adj Close, PEP)	0	0.055847	0.120965
(Adj Close, KO)	0	0.048256	0.120660
(Adj Close, JNJ)	0	0.054099	0.117171
(Adj Close, TM)	0	0.042555	0.147653

(continued)

Table 5-1. *(continued)*

Symbol	Cluster	Returns	Volatility
(Adj Close, HMC)	2	0.010742	0.167794
(Adj Close, MSBHY)	2	0.019532	0.171736
(Adj Close, SNE)	3	0.074899	0.225855
(Adj Close, XOM)	2	-0.003797	0.159237
(Adj Close, CVX)	2	0.029708	0.184557
(Adj Close, VLO)	3	0.086470	0.263032
(Adj Close, F)	2	0.028089	0.216344
(Adj Close, BAC)	3	0.053830	0.244936

To understand the k-means performance, we resort to the silhouette method, which examines the mean intracluster and the mean near-cluster distance for each sample. The value obtained is considered the silhouette score; it measures the separation. A silhouette score ranges from -1 to 1. Specifically, -1 shows poor model performance, and 1 shows optimal model performance. Listing 5-7 finds the score.

Listing 5-7. Find the Silhouette Score

```
from sklearn import metrics
y_predkmeans = pd.DataFrame(kmeans.predict(X))
y_predkmeans = y_predkmeans.dropna()
metrics.silhouette_score(X,y_predkmeans)
0.4260002825147118
```

The silhouette score is 0.42. The score suggests that the model does not sufficiently interpret the data.

Conclusions

This chapter introduced an unsupervised learning model that helps investors better manage risk and select a group of best-performing assets. We used the k-means model to assign data points to distinct clusters. We used the silhouette score to evaluate the performance of the model. After careful consideration, we found that the model shows the characteristics of a well-behaved cluster model. The silhouette score is closer to 1 than -1. However, there is room for improvement. There are overlaps in some clusters, but they are not excessively large enough to affect the conclusions.

CHAPTER 6

Future Price Prediction Using Linear Regression

This chapter introduces the parametric method, also called the *linear regression method*. We use this method to determine the nature of the relationship between an independent variable (continuous or categorical) and a dependent variable (always continuous). Whereas independent variables are continuous variables or categorical variables, a dependent variable is inevitably a continuous variable. It investigates how a change in an independent variable influences a change in a dependent variable. The most conventional model for finding estimates of an intercept and a slope is the least squared model. We express the formula as shown in Equation 6-1.

$$\hat{y} = \beta_0 + \beta_1 X_1 + \varepsilon_i \qquad \text{(Equation 6-1)}$$

Here, $\hat{y}$ represents an expected dependent variable, β_0 represents an intercept, β_1 represents a slope, X_1 represents an independent variable, and ε_i represents the error terms (the residual for the ith of n data points), expressed as shown in Equation 6-2.

$$e_i = y_i - \widehat{y}_i \qquad \text{(Equation 6-2)}$$

© Tshepo Chris Nokeri 2021
T. C. Nokeri, *Implementing Machine Learning for Finance*,
https://doi.org/10.1007/978-1-4842-7110-0_6

The least squares model ensures that the sum squares of residuals are small. We find the sum squares of residuals by applying the property in Equation 6-3.

$$e_1^2 + e_2^2 \ldots e_i^2 \qquad \text{(Equation 6-3)}$$

This property assumes that residuals are always equal to zero and estimates are unbiased.

Linear Regression in Practice

In this chapter, we predict the closing price of gold based on the highest and lowest price changes and returns. Listing 6-1 scraps data from Yahoo Finance.

Listing 6-1. Scraped Data

```
from pandas_datareader import data
start_date = '2010-11-01'
end_date = '2020-11-01'
ticker = 'GC=F'
df = data.get_data_yahoo(ticker, start_date, end_date)
df_orig=df
```

Listing 6-2 calculates the highest and lowest price changes and the returns of the stock price (see Table 6-1).

Listing 6-2. Calculate the Highest and Lowest Price Change and Returns

```
df['HL_PCT']=(df['High']-df['Low'])/df['Adj Close'] *100.0
df['PCT_change']= (df['Adj Close']-df['Open'])/df['Open']
*100.0
```

```
df = df[['Adj Close','HL_PCT','PCT_change','Volume']]
date = df.index
df.head()
```

Table 6-1. *Dataset*

Date	Adj Close	HL_PCT	PCT_change	Volume
2010-11-01	1350.199951	0.711013	-0.742490	40.0
2010-11-02	1356.400024	0.095846	-0.029483	17.0
2010-11-03	1337.099976	2.370799	-1.058166	135.0
2010-11-04	1382.699951	1.728504	1.030248	109.0
2010-11-05	1397.300049	1.417022	0.474588	109.0

Correlation Methods

Correlation estimates the apparent strength of a linear relationship between an independent variable and a dependent variable. Prior to training a regression model, we must identify the severity of the association between variables since the degree of severity influences the performance of a model. There are three principal correlation methods for determining the correlation among variables: the Pearson correlation method, which estimates the correlation among continuous variables; the Kendall correlation method, which estimates the association between rankings and ordinal variables; and the Spearman correlation method, which also estimates the association between rankings of a combination of variables.

The Pearson Correlation Method

Given that we are dealing with continuous variables, we use the Pearson correlation method, which produces values that range from -1 to 1. Here, -1 shows a strong negative correlation relationship, 0 shows no correlation relationship, and 1 shows a strong positive correlation relationship. Listing 6-3 produces the Pearson correlation matrix. We use a heatmap to visually represent the matrix (see Figure 6-1). To install seaborn in the conda environment, use conda install -c anaconda seaborn.

Listing 6-3. Pearson Correlation Matrix

```
import seaborn as sns
dfcorr = df.corr(method="pearson")
sns.heatmap(dfcorr, annot=True,annot_kws={"size":12},cmap="cool
warm")
plt.show()
```

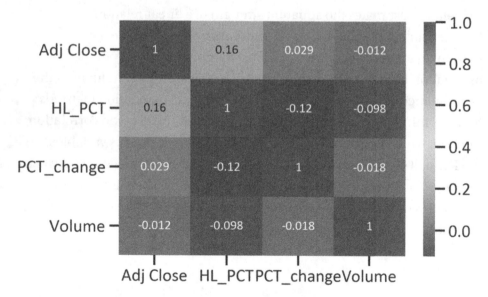

Figure 6-1. *Pearson correlation matrix heatmap*

There is a line of 1s that goes from the top left to the bottom right. This means that each variable perfectly correlates with itself. The correlation is 0.98 (which is close to 1).

The Covariance Method

Listing 6-4 estimates joint variability between the two variables. It estimates how variables vary together (see Figure 6-2).

Listing 6-4. Estimate and Plot a Covariance Matrix

```
dfcov =df.cov()
sns.heatmap(dfcov, annot=True,annot_kws={"size":12},cmap=
"coolwarm")
plt.show()
```

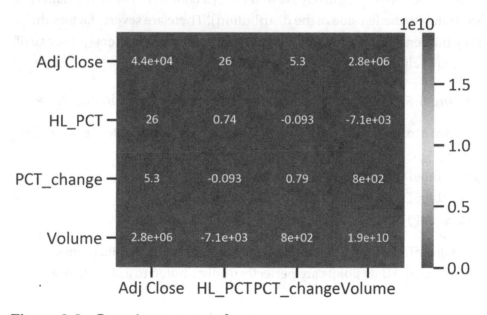

Figure 6-2. *Covariance matrix heatmap*

The covariance matrix confirms more positive joint variability. When we use the Pearson correlation method, we have to first find the covariance since the coefficients of the method are the covariance between two variables divided by their deviation away from the mean value.

Pairwise Scatter Plots

A pairwise scatter plot is typically needed to test for normality. We consider data as normally distributed when independent observations are near the mean value. Regression models assume normality: non-normal data results in negative model performance. In cases where the data is not normal, we may perform a data transformation, i.e., square root transformation for positively skewed data (a condition where the data is saturated to the right side of the distribution) or exponential/power transformation for negatively skewed data (a condition where the data is saturated to the left side of the distribution). There are several factors that may influence non-normality, i.e., missing values and outliers (presence of extreme values), among others. See Listing 6-5.

Listing 6-5. Highest and Lowest Price Changes and Closing Price

```
sns.jointplot(x="HL_PCT",y="Adj Close",data=df,kind="reg",color
="navy")
plt.ylabel("Adj Close")
plt.xlabel("HL_PCT")
plt.show()
```

Figure 6-3 shows a straight line that cuts through the data points. Moreover, the data points are perfectly undistributed to a straight line.

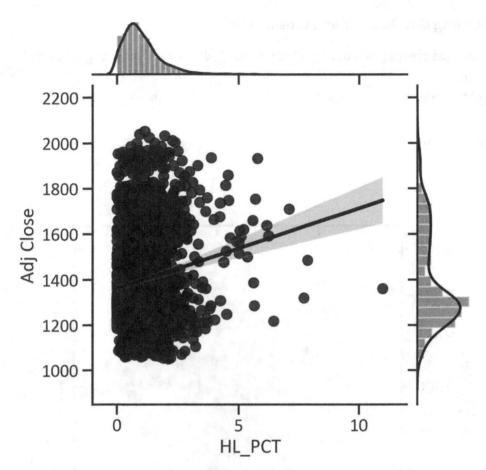

Figure 6-3. *Highest and lowest price changes and closing price scatter plot*

Listing 6-6 constructs a pairwise scatter plot that shows the correlation between the returns and the closing price (see Figure 6-4). The joint plot also shows that HL_PCT is negatively skewed and that Adj Close is positively skewed.

Listing 6-6. Returns and Closing Price

```
sns.jointplot(x="Corr",y="Adj Close",data=xy,kind="reg",color="
red")
plt.ylabel("Adj Close")
plt.xlabel("Corr")
plt.show()
```

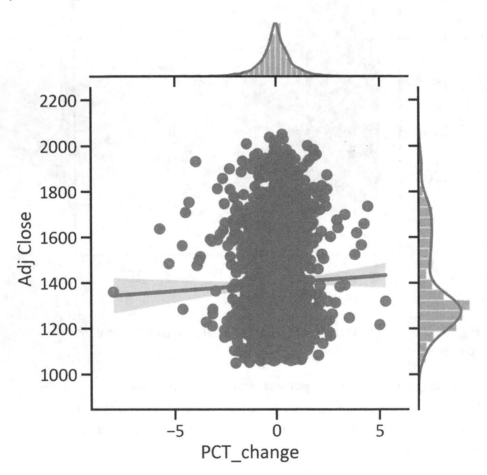

Figure 6-4. *Returns and closing price scatter plot*

We can develop a formula for a straight-line relationship. Moreover, the straight line is not linear. Figure 6-5 illustrates the association between the volume and the adjusted closing price. The figure also shows that the PCT_change follows a normal distribution (the distribution is bell-shaped). See Listing 6-7.

Listing 6-7. Volume and Closing Price

```
sns.jointplot(x="Corr",y="Adj
Close",data=xy,kind="reg",color="red")
plt.ylabel("Adj Close")
plt.xlabel("Corr")
plt.show()
```

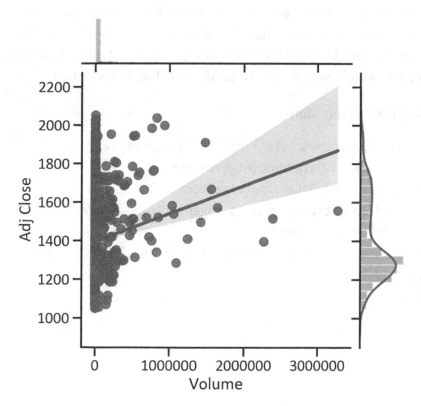

Figure 6-5. *Volume and closing price scatter plot*

Figure 6-5 shows a linear straight line, but the data points are not close to the line.

Eigen Matrix

In the previous section we discovered the strength of the linear relationship among variables. Next, we must diagnose the severity of correlation among the variables using eigenvalues. The eigenvalue represents a flattened variance of the correlation matrix. A combination of more than two variables highly correlated may reduce the predicted power of the model. An eigen matrix is a convenient tool in model selection. Table 6-2 shows an equivalent number of eigenvalues as the variables. An eigenvalue less than 0 indicates no multicollinearity; between 10 and 100 shows moderate multicollinearity, and over 100 shows extreme multicollinearity. Listing 6-8 returns the eigen matrix. Multicollinearity is a problem in which more than two variables are highly correlated.

Listing 6-8. Eigen Matrix

```
eigenvalues, eigenvectors = np.linalg.eig(dfcov)
eigenvalues = pd.DataFrame(eigenvalues)
eigenvectors = pd.DataFrame(eigenvectors)
eigen = pd.concat([eigenvalues,eigenvectors],axis=1)
eigen.index = df.columns
eigen.columns = ("Eigen values","Adj Close","HL_PCT","PCT_
change","Volume")
eigen
```

Table 6-2. *Eigen Matrix*

	Eigen values	Adj Close	HL_PCT	PCT_change	Volume
Adj Close	1.921528e+10	-1.448570e-04	1.000000	-5.758481e-04	-2.688695e-04
HL_PCT	4.334201e+04	3.710674e-07	0.000624	8.106354e-01	5.855510e-01
PCT_change	6.528812e-01	-4.166475e-08	0.000119	5.855510e-01	-8.106356e-01
Volume	8.540129e-01	-1.000000e+00	-0.000145	3.598192e-07	2.900014e-07

Table 6-2 highlights that there is moderate multicollinearity (the eigenvalue is less than 10).

Further Descriptive Statistics

There are several ways to summarize the data. Listing 6-9 summarizes the opening and closing prices across time (see Figure 6-6).

Listing 6-9. Opening and Closing Prices

```
df_orig[["Open","Close"]].head(20).plot(kind='bar',cmap="rainbow")
plt.ylabel("Price")
plt.show()
```

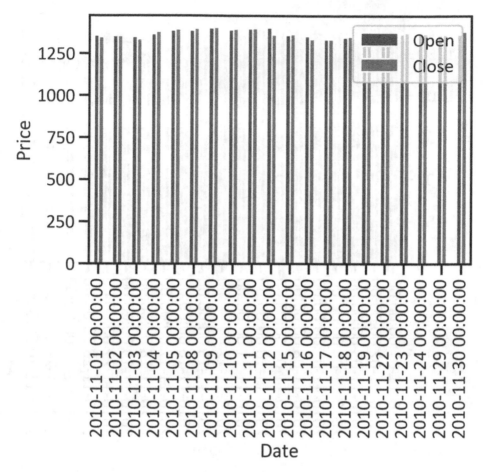

Figure 6-6. *Opening and closing prices*

Figure 6-6 shows that there are no major deviations of the closing price away from the opening price. Figure 6-7 depicts the low price and closing price across time. See Listing 6-10.

Listing 6-10. Low and Closing Prices

```
df_orig[["Low","Close"]].head(20).plot(kind="bar",cmap="rainbow")
plt.ylabel("Price")
plt.show()
```

113

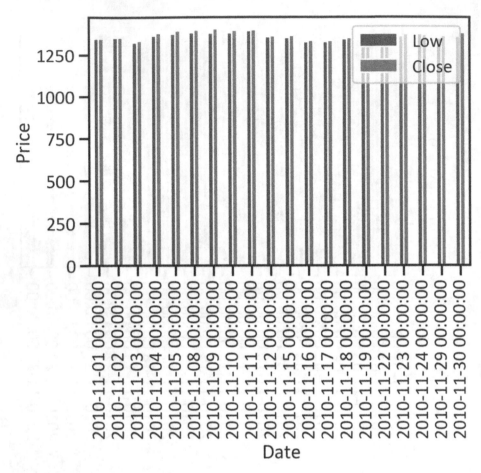

Figure 6-7. *Low and closing prices*

To further understand the changes in prices, we can graphically represent the high price and closing price across time (see Figure 6-8). See Listing 6-11.

Listing 6-11. High and Adjusted Close Prices

```
df_orig[['High','Close']].head(20).plot(kind='bar',cmap="rainbow")
plt.ylabel("Price")
plt.show()
```

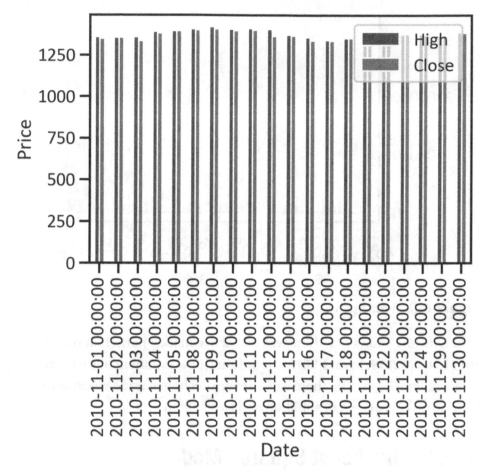

Figure 6-8. *Low and adjusted close prices*

Listing 6-12 shows the volume of trades over time from November 1, 2020, to November 2, 2020 (see Figure 6-9).

Listing 6-12. Volume

```
df_orig.Volume.plot(color="green")
plt.ylabel("Volume")
```

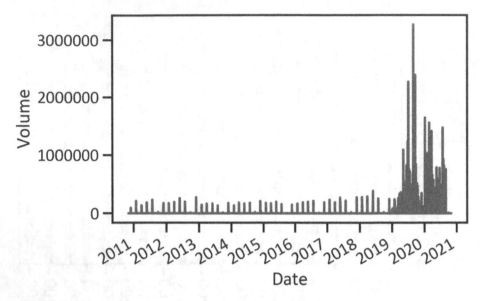

Figure 6-9. *Volume*

Figure 6-1 shows that there was less active market participation from 2011 to early 2019. In mid-2019, the volume grew exponentially to over three million. Post-2019, there was a severe decline in trade volume; the volume was between half a million and two million.

Develop the Least Squares Model

Listing 6-13 splits the data into training and test data applying the 80/20 split ratio.

Listing 6-13. Split Data into Training and Test Data

```
from sklearn.model_selection import train_test_split
x_train, x_test, y_train, y_test = train_test_split(x,y,test_
size=0.2, random_state=0)
```

Listing 6-14 normalizes the training data using the `StandardScaler()` method (transforms data in such a way that the mean value is 0 and the standard deviation is 1).

Listing 6-14. Normalize Data

```
from sklearn.model_selection import StandardScaler
scaler = StandardScaler()
x_train = scaler.fit_transform(x_train)
x_test = scaler.transform(x_test)
```

Listing 6-15 fits the least squares model with default hyperparameters.

Listing 6-15. Develop the Least Squares Model

```
from sklearn.linear_model import LinearRegression
lm = LinearRegression()
lm.fit(x_train,y_train)
```

Listing 6-16 defines a function to find the mean value and standard deviation of the cross-validation score by applying the R^2 as a criterion for finding the cross-validation scores.

Listing 6-16. Develop a Function to Obtain Cross-Validation Mean and Standard Deviation

```
from sklearn.model_selection import cross_val_score
def get_val_score(model):
    scores = cross_val_score(model, x_train, y_train,
    scoring="r2")
    print("CV mean: ", np.mean(scores))
    print("CV std: ", np.std(scores))
    print("\n")
```

Listing 6-17 prints the mean value and standard deviation value of the cross-validation scores.

Listing 6-17. Cross-Validation Mean and Standard Deviation

```
get_val_score(lm)
CV mean:   0.9473235769188586
CV std:   0.018455084710127526
```

Listing 6-18 finds the names and default values of a parameter.

Listing 6-18. Find Default Parameters

```
lm.get_params()
```

Listing 6-19 creates a grid model. A hyperparameter represents settings or values that we must configure for a model prior to training. We perform hyperparameter optimization to identity values that produce optimal model performance. The GridSearchCV method considers all parameters and discovers the most suitable combination. For example, in Listing 6-19 we want to find whether we must fit the intercept, normalize X prior to fitting the model, and copy X.

Listing 6-19. Develop a Grid Model

```
from sklearn.model_selection import GridSearchCV
param_grid = {'fit_intercept':[True,False],
              'normalize':[True,False],
              'copy_X':[True, False]}
grid_model   = GridSearchCV(estimator=lm,
                            param_grid=param_grid,
                            n_jobs=-1)
grid_model.fit(x_train,y_train)
```

Listing 6-20 finds the best score and best hyperparameters.

Listing 6-20. Hyperparameter Optimization

```
print("Best score: ", grid_model.best_score_, "Best parameters:
", grid_model.best_params_)
```

Here's the result:

- *Best score*: 0.9473235769188586

- *Best parameters*: {'copy_X': True, 'fit_intercept': True,
  'normalize': False}

Listing 6-21 completes the least squares model by applying the hyperparameters returned by the grid model to train the model.

Listing 6-21. Finalize the Least Squares Model

```
lm = LinearRegression(copy_X= True,
                      fit_intercept= True,
                      normalize= False)
lm.fit(x_train,y_train)
```

Listing 6-22 finds the intercept. An intercept is the mean value of an independent variable, given that we hold a dependent variable constant.

Listing 6-22. Intercept

```
lm.intercept_
15.725513886138613
```

Listing 6-23 estimates the coefficients.

Listing 6-23. Coefficients

```
lm.coef_
array([1.45904887, 0.04329147])
```

Model Evaluation

Listing 6-24 applies the predict() method to return Table 6-3 (it highlights the values that the regressor produces).

Listing 6-24. Forecast Values

```
y_pred = lm.predict(x_test)
pd.DataFrame(y_pred,columns=["Forecast"])
```

Table 6-3. *Forecast*

	Forecast
0	1469.400024
1	1466.699951
2	1475.599976
3	1478.400024
4	1475.000000
...	...
253	1902.699951
254	1908.800049
255	1876.199951
256	1865.599976
257	1877.400024

Listing 6-25 plots future instances of the adjusted close price (see Figure 6-10).

Listing 6-25. Forecast

```
num_samples = df.shape[0]
df['Forecast'] = np.nan
df['Forecast'][int(0.9*num_samples):num_samples]=y_pred
df['Adj Close'].plot(color="navy")
df['Forecast'].plot(color="red")
plt.legend(loc="best")
plt.xlabel('Date')
plt.ylabel('Price')
plt.show()
```

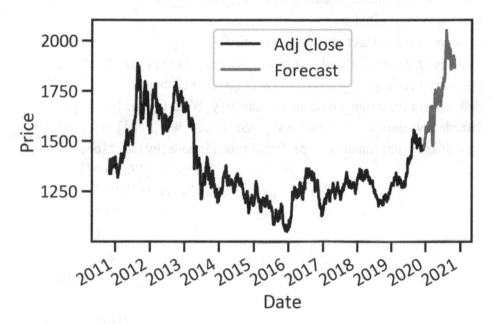

Figure 6-10. *Forecast*

Listing 6-26 returns a table that contains information about the performance of the regressor (see Table 6-4). It shows key regression evaluation metrics, such as the mean absolute error (the magnitude of errors without considering the direction), mean squared error (the variability explained after considering a linear relationship, and R2 score (the variability explained by the model about the data).

Listing 6-26. Model Performance

```
from sklearn import metrics
MAE = metrics.mean_absolute_error(y_test,y_pred)
MSE = metrics.mean_squared_error(y_test,y_pred)
RMSE = np.sqrt(MSE)
R2 = metrics.r2_score(y_test,y_pred)
EV = metrics.explained_variance_score(y_test,y_pred)
MGD = metrics.mean_gamma_deviance(y_test,y_pred)
MPD = metrics.mean_poisson_deviance(y_test,y_pred)
lmmodelevaluation = [[MAE,MSE,RMSE,R2,EV,MGD,MPD]]
lmmodelevaluationdata = pd.DataFrame(lmmodelevaluation,
                                index = ["Values"],
                                columns = ["MAE",
                                            "MSE",
                                            "RMSE",
                                            "R2",
                                            "Explained
                                            variance
                                            score",
                                            "Mean gamma
                                            deviance",
                                            "Mean Poisson
                                            deviance"]).
                                            transpose()

lmmodelevaluationdata
```

Table 6-4. *Model Performance*

	Values
MAE	2.820139e-13
MSE	1.150198e-25
RMSE	3.391457e-13
R2	1.000000e+00
Explained variance score	1.000000e+00
Mean gamma deviance	-2.581914e-18
Mean Poisson deviance	0.000000e+00

Table 6-4 highlights that the model explains 100 percent of the variability in the data. There is a significant correlation in the time-series data. On average, the magnitude of errors without considering the direction is 2.82, and the mean sum of errors is 1.15.

Conclusion

In this chapter, we briefly covered the least squared model and its application. To begin with, we covered covariance and correlation. Next, we showed you how to design, build, and test a regressor. After carefully reviewing the regressor's performance, we found that the model best explains the data. We may use it for reliable future price predictions. In the subsequent chapter, we cover market simulation.

CHAPTER 7

Stock Market Simulation

The stock exchange market typically responds to socio-economic conditions. The active participation of key market players undoubtedly creates liquidity in the market. There are other factors that influence variations in the market besides the large transactions that key market makers make. For instance, markets react to news of social events, economic events, natural disasters, pandemics, etc. In certain conditions, the effects of those events drive price movement drastically. Systematic investors (investors who base their trade execution on a system that depends on quantitative models) need to prepare for future occurrences to preserve capital by learning from near-real-life conditions. If you inspect industries that involve considerable risk, for example, aerospace and defense, you will notice that the subjects learn through simulation. Simulation is a method that involves generating conditions that mimic the actual world so that subjects know how to act and react in a condition similar to preceding ones. In finance, we deal with large funds. They habitually take risk into account when experimenting and testing models.

In this chapter, we implement a Monte Carlo simulation to simulate considerable changes in the market without exposing ourselves to risk. Simulating the market helps identify patterns in preceding occurrences and forecasts future prices with reasonable certainty. If we can reconstruct the actual world, then we can understand the preceding market behavior and

© Tshepo Chris Nokeri 2021
T. C. Nokeri, *Implementing Machine Learning for Finance*,
https://doi.org/10.1007/978-1-4842-7110-0_7

predict future market behavior. In the preceding chapters, we sufficiently covered models for sequential pattern identification and forecasting. We use the `panda_montecarlo` framework to perform the simulation. To install it in the Python environment, use `pip install pandas-montecarlo`.

When an investor trades a stock, they expect their speculation to yield returns over a certain period. Unexpected events occur, and they can influence the direction of prices. To build sustained profitability over the long run, investors can develop models to recognize the underlying pattern of preceding occurrences and to forecast future occurrences.

Let's assume you are a prospective conservative investor with an initial investment capital of $5 million US dollars. After conducting a comprehensive study, you have singled out a set of stocks to invest in. You can use Monte Carlo simulation to further understand the risks of investing in those assets.

Understanding Value at Risk

The most convenient way to estimate risk is by applying the value at risk (VAR). It reveals the degree of financial risk we are exposing an investment portfolio to. It shows the minimum capital required to compensate for losses at a specified probability level. There are two primary ways of estimating VAR; we can either use the variance-covariance method or simulate it by applying Monte Carlo.

Estimate VAR by Applying the Variance-Covariance Method

The variance-covariance method makes strong assumptions about the structure of the data. The method assumes the underlying structure of the data is linear and normal. It is also sensitive to missing data, nonlinearity, and outliers. To estimate the standard VAR, we first find the returns, and then we create the covariance matrix and find the mean value and standard deviation of the investment portfolio. Thereafter, we estimate the inverse of the normal cumulative distribution and compute the VAR. Figure 7-1 shows the VAR of a portfolio alongside the simulated distributions.

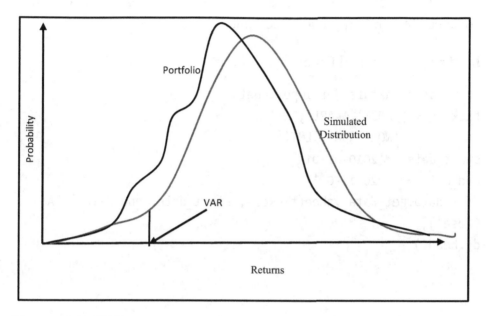

Figure 7-1. *VAR*

Here we show you how to calculate the VAR by applying the variance/covariance calculation.

Assume:

Investment capital is $5,000,000.

Standard deviation from an annual trading calendar (252 days) is 9 percent.

Using the z-score (1.65) at 95 percent confidence
interval, the value at risk is as follows:

$5,000,0000*1.645*.09 = $740 250

Listing 7-1 scrapes the stock prices for Amazon, Apple, Walgreens
Boots Alliance, Northrop Grumman Corporation, Boeing Company, and
Lockheed Martin Corporation (see Table 7-1).

Listing 7-1. Scraped Data

```
from pandas_datareader import data
tickers = ['AMZN','AAPL','WBA',
           'NOC','BA','LMT']
start_date = '2010-01-01'
end_date = '2020-11-01'
df = data.get_data_yahoo(tickers, start_date, end_date)[['Adj
Close']]
df.head()
```

Table 7-1. Dataset

Attributes	Adj Close					
Symbols	AMZN	AAPL	WBA	NOC	BA	LMT
Date						
2010-01-04	133.899994	6.539882	28.798639	40.206833	43.441975	53.633926
2010-01-05	134.690002	6.551187	28.567017	40.277546	44.864773	54.192230
2010-01-06	132.250000	6.446983	28.350834	40.433144	46.225727	53.396633
2010-01-07	130.000000	6.435065	28.520689	40.850418	48.097031	51.931023
2010-01-08	133.520004	6.477847	28.559305	40.624100	47.633064	52.768509

Listing 7-2 specifies the investment weights of the portfolio.

Listing 7-2. Specify Investment Weights

```
weights = np.array([.25, .3, .15, .10, .24, .7])
```

From then on we specify the initial investment capital in the portfolio. See Listing 7-3.

Listing 7-3. Specify Initial Investment

```
initial_investment = 5000000
```

Listing 7-4 estimates the daily returns (see Table 7-2).

Listing 7-4. Estimate Daily Returns

```
returns = df.pct_change()
returns.tail()
```

Table 7-2. Daily Returns

Attributes	Adj Close					
Symbols	AMZN	AAPL	WBA	NOC	BA	LMT
Date						
2020-10-26	0.000824	0.000087	-0.021819	0.004572	-0.039018	-0.015441
2020-10-27	0.024724	0.013472	-0.032518	-0.024916	-0.034757	-0.016606
2020-10-28	-0.037595	-0.046312	-0.039167	-0.028002	-0.045736	-0.031923
2020-10-29	0.015249	0.037050	-0.030934	-0.004325	0.001013	0.004503
2020-10-30	-0.054456	-0.056018	0.015513	-0.008790	-0.026300	-0.006554

Table 7-2 highlights the first five rows of each stock's daily return. Listing 7-5 estimates the joint variability between the stocks in the portfolio (see Table 7-3).

Listing 7-5. Covariance Matrix

```
cov_matrix = returns.cov()
cov_matrix
```

Table 7-3. Covariance Matrix

Attributes		Adj Close					
Adj Close	Symbols	AMZN	AAPL	WBA	NOC	BA	LMT
	AMZN	0.000401	0.000159	0.000096	0.000092	0.000135	0.000081
	AAPL	0.000159	0.000318	0.000098	0.000098	0.000161	0.000093
	WBA	0.000096	0.000098	0.000303	0.000097	0.000131	0.000088
	NOC	0.000092	0.000098	0.000097	0.000204	0.000165	0.000150
	BA	0.000135	0.000161	0.000131	0.000165	0.000486	0.000162
	LMT	0.000081	0.000093	0.000088	0.000150	0.000162	0.000174

Listing 7-6 estimates the VAR. First, we specify the average daily returns, then we specify the confidence interval and cutoff value, and finally we obtain the mean and standard deviations. Subsequently, we find the inverse of the distribution.

Listing 7-6. Estimate Value at Risk

```
conf_level1 = 0.05
avg_rets = returns.mean()
port_mean = avg_rets.dot(weights)
port_stdev = np.sqrt(weights.T.dot(cov_matrix).dot(weights))
mean_investment = (1+port_mean) * initial_investment
stdev_investment = initial_investment * port_stdev
cutoff1 = norm.ppf(conf_level1, mean_investment, stdev_
investment)
var_1d1 = initial_investment - cutoff1
var_1d1
166330.5512926411
```

At a 95 percent confidence interval, the investment portfolio of $5,000,000 will not exceed losses greater than 166,330.55 a day. Listing 7-7 prints the 10-day VAR.

Listing 7-7. Print 10-Day VAR

```
var_arry = []
num_days = int(10)
for x in range(1, num_days+1):
    var_array.append(np.round(var_1d1 * np.sqrt(x),2))
    print(str(x) + " day VaR @ 95% confidence: " + str
    (np.round(var_1d1 * np.sqrt(x),2)))
1 day VaR @ 95% confidence: 166330.55
2 day VaR @ 95% confidence: 235226.92
3 day VaR @ 95% confidence: 288092.97
```

```
4 day VaR @ 95% confidence: 332661.1
5 day VaR @ 95% confidence: 371926.42
6 day VaR @ 95% confidence: 407424.98
7 day VaR @ 95% confidence: 440069.27
8 day VaR @ 95% confidence: 470453.84
9 day VaR @ 95% confidence: 498991.65
10 day VaR @ 95% confidence: 525983.39
```

Listing 7-8 graphically represents the VAR over a 10-day period (see Figure 7-2).

Listing 7-8. 10-Day VAR

```
plt.plot(var_array, color="navy")
plt.xlabel("No. of Days")
plt.ylabel("MaX Portfolio Loss (USD)")
plt.show()
```

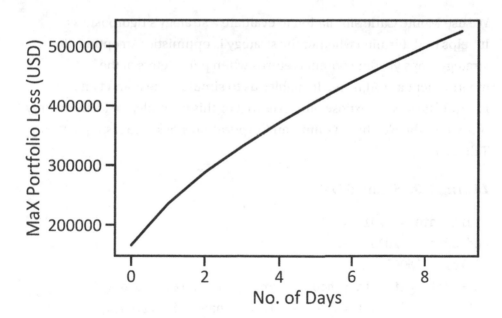

Figure 7-2. *Max portfolio loss (VAR) over a 15-day period*

Figure 7-2 shows that as we increase the number of days, the maximum investment portfolio increases too. In the next section, we explored Monte Carlo simulation.

Understanding Monte Carlo

Monte Carlo simulation uses resampling techniques to combat sequential problems. It reconstructs real-life conditions to identify and understand the results of the preceding occurrences and predict future occurrences. It also enables us to experiment with different investment strategies. Not only that, but it performs repetitive measurements on random variables that come from a normal distribution to determine the probability of each output, and then it assigns a confidence interval output.

Application of Monte Carlo Simulation in Finance

We use Monte Carlo simulation to evaluate a strategy's rigorousness. It helps us determine whether the strategy is optimistic. An optimistic strategy stops yielding optimal returns when parameters of the environment are adjusted. It enables us to simulate the market and identify the risk we expose ourselves to. For this example, we focus on only one stock, the Northrop Grumman Corporation stock. See Listing 7-9 and Table 7-4.

Listing 7-9. Scraped Data

```
start_date = '2010-11-01'
end_date = '2020-11-01'
ticker = 'NOC'
df = data.get_data_yahoo(ticker, start_date, end_date)
df['return'] = df['Adj Close'].pct_change().fillna(0)
df.head()
```

Table 7-4. Dataset

Date	High	Low	Open	Close	Volume	Adj Close	return
2010-11-01	58.257679	56.974709	57.245762	57.381287	1736400.0	46.094521	0.000000
2010-11-02	58.438381	57.833035	57.833035	58.357063	1724000.0	46.878361	0.017005
2010-11-03	58.501625	57.354179	58.221539	58.076981	1598400.0	46.653362	-0.004800
2010-11-04	59.143108	58.140224	58.483555	58.971443	2213600.0	47.371891	0.015401
2010-11-05	59.260567	58.826885	58.926270	59.034691	1043900.0	47.422695	0.001072

Run Monte Carlo Simulation

Listing 7-10 applies the panda_montecarlo() method to run the Monte Carlo simulation with five simulations. Also, we set the bust/max drawdown to -10.0 percent and the goal threshold to +100.0 percent.

Listing 7-10. Run the Monte Carlo Simulation

```
mc = df['return'].montecarlo(sims=10, bust=-0.1, goal=1)
```

Plot Simulations

Listing 7-11 plots the 10 simulations that the Monte Carlo simulation ran (see Figure 7-3).

Listing 7-11. Simulations

```
mc.plot(title="")
```

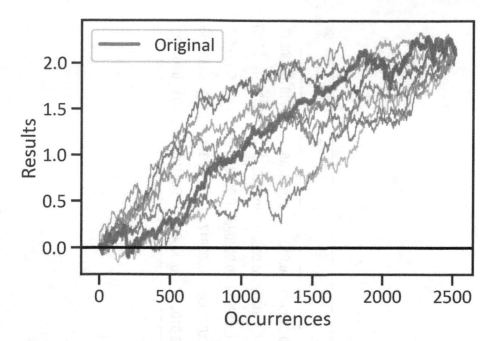

Figure 7-3. *Monte Carlo simulations*

Figure 7-3 shows the simulation results. It highlights a dominant upward trend. Listing 7-12 tabulates raw simulations (see Table 7-5).

Listing 7-12. Raw Simulations

```
pd.DataFrame(mc.data).head()
```

Table 7-5. *Raw Simulations*

	Original	1	2	3	4	5	6	7	8	9
0	0.000000	-0.018168	-0.002659	-0.021098	-0.006357	0.018090	-0.010110	-0.006555	-0.004119	0.020557
1	0.017005	0.005521	-0.014646	0.000541	-0.009134	-0.000602	-0.006801	-0.003373	0.004342	-0.002257
2	-0.004800	-0.003900	-0.002494	0.027738	0.005924	-0.012834	-0.004095	-0.003315	0.000116	0.113851
3	0.015401	0.006688	0.001144	0.005586	-0.004924	0.011399	0.009817	-0.001273	0.006164	0.023355
4	0.001072	0.005617	0.010052	0.011748	0.007878	0.010080	-0.008666	-0.005126	0.036459	0.015438

Listing 7-13 returns the statistics of the simulation model (Table 7-6).

Listing 7-13. Monte Carlo Statistics

```
ev = pd.DataFrame(mc.stats, index=["s"])
ev
```

Table 7-6. *Monte Carlo Statistics*

	min	max	mean	median	std	maxdd	bust	goal
s	2.094128	2.094128	2.094128	2.094128	2.145155e-15	-0.164648	0.2	0.8

Table 7-6 highlights the dispersion. It also highlights the maximum drawdown (the maximum amount of loss from a peak).

Conclusions

When investing in a stock or a set of stocks, it is important to understand, quantify, and mitigate the underlying risk in an investment portfolio. This chapter discussed VAR; thereafter, it showed ways to calculate the VAR of an investment portfolio, followed by simulating stock market changes by applying Monte Carlo simulation. The simulation technique has many applications; we can also use it for asset price structuring, etc. The next chapter further expands on the investment portfolio and risk analysis.

CHAPTER 8

Market Trend Classification Using ML and DL

Thus far, we have progressively introduced the parametric method. All the problems we solved involved a continuous variable. We can also make use of the *nonparametric method* to predict the possible direction of the market. This chapter presents the nonparametric (or nonlinear) method, also called the *classification method*. This prevalent method operates on independent variables and triggers a bounded value. It is suitable when dealing with a categorical dependent variable (a dependent variable that is limited by a specific range). There are two primary types of classification methods: the binary classification method, which is used when the dependent variable has two outcomes, and the multiclass classification method, which is used when the dependent variable is a categorical variable with over two outcomes. In this chapter, we use both SciKit-Learn and Keras. The SciKit-Learn library is pre-installed in the Python environment. To install Keras in the Python environment, we use `pip install Keras`, and in the conda environment, we use `conda install -c conda-forge keras`.

© Tshepo Chris Nokeri 2021
T. C. Nokeri, *Implementing Machine Learning for Finance*,
https://doi.org/10.1007/978-1-4842-7110-0_8

Classification in Practice

Recognizing supply-and-demand activities helps investors make well-informed investment decisions. In this chapter, we create a categorical variable with the two outcomes 0 and 1, where 0 represents a downward market and 1 represents an upward market. We use a logistic classifier to predict future classes. Listing 8-1 applies the get_data_yahoo() method to get the price data of crude oil[1] from November 1, 2010, to November 1, 2010 (see Table 8-1).

Listing 8-1. Scraped Data

```
from pandas_datareader import data
start_date = '2010-11-01'
end_date = '2020-11-01'
ticker = 'CL=F'
df = data.get_data_yahoo(ticker, start_date, end_date,)
df.head()
```

---

[1]https://finance.yahoo.com/quote/CL=F/

Table 8-1. Dataset

Date	High	Low	Open	Close	Volume	Adj Close
2010-11-01	83.860001	81.320000	81.449997	82.949997	358535.0	82.949997
2010-11-02	84.470001	82.830002	82.879997	83.900002	281834.0	83.900002
2010-11-03	85.360001	83.570000	84.370003	84.690002	393735.0	84.690002
2010-11-04	86.830002	84.919998	85.089996	86.489998	317997.0	86.489998
2010-11-05	87.430000	85.959999	86.599998	86.849998	317997.0	86.849998

Listing 8-2 estimates the returns and log returns (see Table 8-2).

Listing 8-2. Estimate Returns and Log Returns

```
df = df.dropna()
df['pct_change'] = df["Adj Close"].pct_change()
df['log_ret'] = np.log(df["Adj Close"]) - np.log(df["Adj
Close"].shift(1))
df.head()
```

Table 8-2. *Dataset with Estimated Returns and Log Returns*

Date	High	Low	Open	Close	Volume	Adj Close	pct_change	log_ret
2010-11-01	83.860001	81.320000	81.449997	82.949997	358535.0	82.949997	NaN	NaN
2010-11-02	84.470001	82.830002	82.879997	83.900002	281834.0	83.900002	0.011453	0.011388
2010-11-03	85.360001	83.570000	84.370003	84.690002	393735.0	84.690002	0.009416	0.009372
2010-11-04	86.830002	84.919998	85.089996	86.489998	317997.0	86.489998	0.021254	0.021031
2010-11-05	87.430000	85.959999	86.599998	86.849998	317997.0	86.849998	0.004162	0.004154

Listing 8-3 drops the missing values and estimates the market direction (see Table 8-3).

Listing 8-3. Drop Missing Values and Estimate Market Direction

```
df = df.dropna()
df['direction'] = np.sign(df['pct_change']).astype(int)
df.head(3)
```

Table 8-3. Dataset with Estimated Returns, Log Returns, and Market Direction

Date	High	Low	Open	Close	Volume	Adj Close	pct_change	log_ret	direction
2010-11-02	84.470001	82.830002	82.879997	83.900002	281834.0	83.900002	0.011453	0.011388	1
2010-11-03	85.360001	83.570000	84.370003	84.690002	393735.0	84.690002	0.009416	0.009372	1
2010-11-04	86.830002	84.919998	85.089996	86.489998	317997.0	86.489998	0.021254	0.021031	1

Data Preprocessing

Listing 8-4 creates a categorical variable with the two outcomes 0 and 1, where 0 represents a downward market and 1 represents an upward market. This enables us to use a classifier that can predict the probability of two classes. We start by defining the number of lags, and then we estimate daily returns and convert the lag returns into binary classes. We also use the GridSearchCV() method to standardize the data in such a way that the mean value is 0 and the standard deviation is 1.

Listing 8-4. Data Preprocessing

```
from sklearn.preprocessing import StandardScaler
df["direction"] = pd.get_dummies(df["direction"])
from sklearn import preprocessing
x=df.iloc[::,5:8]
y=df.iloc[::,-1]
scaler = StandardScaler()
x=scaler.fit_transform(x)
```

Listing 8-5 splits the data into training data and test data using the 80/20 split ratio.

Listing 8-5. Split Data into Training and Test Data

```
from sklearn.model_selection import train_test_split
x_train, x_test, y_train, y_test =train_test_split(x , y, test_
size=0.2,shuffle= False)
```

Logistic Regression

Although the term *logistic regression* contains the word *regression*, it is not a regression model but a classification model. A linear regression model estimates a continuous variable. Meanwhile, a logistic regression model estimates a categorical dependent variable. Regression models assume the data is linear and comes from a normal distribution, but logistic classifiers are free from those assumptions. In logistic regression, we fit an S-shaped curve (or logistic curve or sigmoid curve) to the data. We use a logistic classifier to predict market movements. Stock prices change significantly in specific periods. If the market should upsurge within a certain period, then there is an upward market or a bullish market. In contrast, if the market price consistently decreases within a specific period, then there is a downward market or bearish market.

Develop the Logistic Classifier

Listing 8-6 finalizes the logistic classifier.

Listing 8-6. Finalize the Logistic Classifier

```
from sklearn.linear_model import LogisticRegression
logreg = LogisticRegression()
logreg.fit(x_train, y_train)
```

Evaluate a Logistic Classifier

Listing 8-7 constructs a table that highlights the predicted market classes (see Table 8-4).

Listing 8-7. Predicted Values

```
y_predlogreg = logreg.predict(x_test)
pd.DataFrame(y_predlogreg,columns=["Forecast"])
```

Table 8-4. *Predicted Classes*

	Forecast
0	1
1	1
2	0
3	1
4	1
...	...
494	1
495	0
496	1
497	0
498	1

Table 8-4 does not provide us with sufficient information about how well the logistic classifier predicts. To find an abstract background of the logistic classifier's performance, we use a confusion matrix.

Confusion Matrix

We typically use a confusion matrix to identify two types of errors: the false positive, which incorrectly predicts that an event took place, and the false negative, which incorrectly predicts an event that never happened. It also highlights the true positive, which correctly predicts an event that took place, and the true negative, which correctly predicts an event that never took place. Listing 8-8 constructs a confusion matrix (see Table 8-5).

Listing 8-8. Confusion Matrix

```
from sklearn import metrics
cmatlogreg = pd.DataFrame(metrics.confusion_matrix(y_test,y_
predlogreg),
                          index=["Actual: Sell","Actual: Buy"],
                          columns=("Predicted:
                          Sell","Predicted: Buy"))
cmatlogreg
```

Table 8-5 does not tell us much besides giving us a count of actual "sell" and actual "buy," and predicted "sell" and predicted "buy." To better understand how the logistic classifier works, we use the classification report.

Table 8-5. *Confusion Matrix Estimates*

	Predicted: Sell	Predicted: Buy
Actual: Sell	234	0
Actual: Buy	0	265

Classification Report

Table 8-8 provides sufficient details about the classifier's performance. It tabulates the accuracy (how often a classifier gets predictions right), precision (how often a classifier is correct), F-1 score (the harmonic mean value of precision and recall), and support (the number of samples of the actual response in that class). It also shows whether there is an imbalance in the data. To understand how this works, in this section we show you how to estimate both accuracy and precision (also refer to Table 8-6).

$$Accuracy = TP + TN\ /\ TP + TN + FP + FN \qquad \text{(Equation 8-1)}$$

$$Precision = TP\ /\ TP + FP \qquad \text{(Equation 8-2)}$$

153

Table 8-6. *Understanding Confusion Matrix Estimates*

Metric	Description
TP	Representing true positives (how many times the classifier predicted a "sell" signal when it was a "sell" signal)
TN	Representing true negatives (how many times the classifier predicted a "buy" signal when it was a "buy" signal)
FP	Representing false positives (how many times the classifier predicted a "sell" signal when it was a "buy" signal)
FP	Representing false negatives (how many times the classifier predicted a "buy" signal when it was a "sell" signal)

To understand how this works, look at Table 8-6. Also refer to Table 8-7.

Table 8-7. *How to Obtain Confusion Matrix Estimates*

	Predicted: Sell	Predicted: Buy
Actual: Sell	TP	FP
Actual: Buy	FN	TN

Listing 8-9 shows the classification report (see Table 8-8).

Listing 8-9. Classification Report

```
creportlogreg =pd.DataFrame(metrics.classification_report(y_
test,y_predlogreg,output_dict=True)).transpose()
creportlogreg
```

Table 8-8. *Classification Report*

	precision	recall	f1-score	support
0	1.0	1.0	1.0	256.0
1	1.0	1.0	1.0	243.0
accuracy	1.0	1.0	1.0	1.0
macro avg	1.0	1.0	1.0	499.0
weighted avg	1.0	1.0	1.0	499.0

Table 8-8 highlights that the logistic classifier is accurate 100 percent of the time (the accuracy is at 1.0). This also shows there is an imbalance in the data. We will not depend on the accuracy score to assess its performance.

ROC Curve

We use the ROC curve to find the area under the curve (AUC). ROC stands for "receiver operating characteristic," which shows the extent to which the classifier distinguishes between classes. The roc_curve( ) method takes the actual classes and probabilities of each class to develop the curve. Listing 8-10 constructs an ROC curve to summarize the trade-offs between the false positive rate and the true positive rate across different thresholds (see Figure 8-1).

Listing 8-10. ROC Curve

```
y_predlogreg_proba = logreg.predict_proba(x_test)[::,1]
fprlogreg, tprlogreg, _ =metrics.roc_curve(y_test,y_predlogreg_
proba)
auclogreg = metrics.roc_auc_score(y_test, y_predlogreg_proba)
plt.plot(fprlogreg, tprlogreg, label="AUC:
"+str(auclogreg),color="navy")
```

```
plt.plot([0,1],[0,1],color="red")
plt.xlim([0.00,1.01])
plt.ylim([0.00,1.01])
plt.xlabel("Specificity")
plt.ylabel("Sensitivity")
plt.legend(loc=4)
plt.show()
```

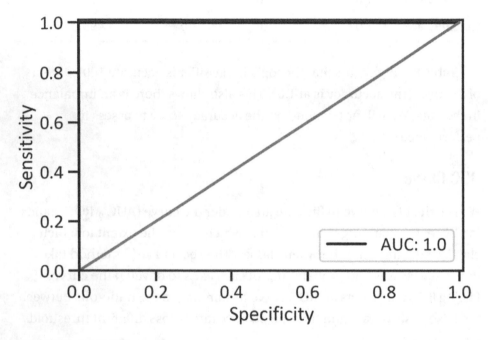

Figure 8-1. *ROC curve*

The AUC score is greater than 0.80. This means the logistic classifier is skillful in distinguishing classes. Ideally, we want a classifier that has an AUC score that is closer to 1.

Learning Curve

Figure 8-2 has two axes: the training set size on the x-axis and the accuracy score on the y-axis. It illustrates how the classifier learns to make accurate predictions as we progressively increase the data. See Listing 8-11.

Listing 8-11. Learning Curve

```
from sklearn.model_selection import learning_curve
trainsizelogreg, trainscorelogreg, testscorelogreg
=learning_curve(logreg, x, y, cv=5, n_jobs=5,train_sizes=np.
linspace(0.1,1.0,50))
trainscorelogreg_mean = np.mean(trainscorelogreg,axis=1)
testscorelogreg_mean = np.mean(testscorelogreg,axis=1)
plt.plot(trainsizelogreg,trainscorelogreg_mean,color="red",labe
l="Training score", alpha=0.8)
plt.plot(trainsizelogreg,testscorelogreg_
mean,color="navy",label="Cross validation score", alpha=0.8)
plt.xlabel("Training set size")
plt.ylabel("Accuracy")
plt.legend(loc=4)
plt.show()
```

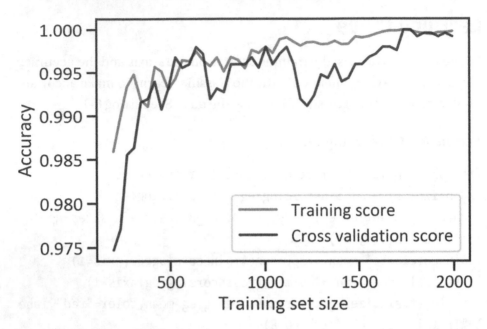

Figure 8-2. *Learning curve*

Figure 8-2 conveys that the classifier commits many mistakes in the foundation phase of the training. The average accuracy score surges as we increase the training set size, and the training score is predominantly beneath the cross-validation score.

Multilayer Layer Perceptron

Chapter 5 covered deep learning and its application in finance. Thereafter, we explained the fundamental structure of an artificial neural network and revealed ways in which you can combat sequential problems using the recurrent neural network (RNN). We developed and assessed the long short-term memory (LSTM) model to forecast future stock prices. In this chapter, we use a multilayer perceptron (MLP) classifier to estimate the probabilities of an upward or downward market. It comprises three layers: the input layer, which receives input values; the hidden layer, which transforms the values; and the output layer, which triggers an output value. The MLP model receives a set of input values, transforms them, and triggers output values in the output layer. It is basically a combination of multiple restricted Boltzmann machines (a neural network that maintains a visible layer and hidden layer). The model addresses the vanishing gradient problem through backward propagation, which involves estimating the gradient from the right to the left (the opposite of forward propagation that involves estimating the gradient from the left to the right).

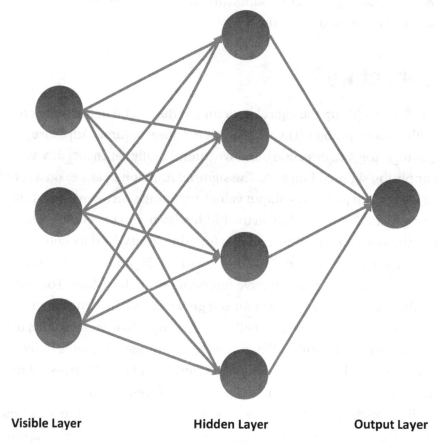

| Visible Layer | Hidden Layer | Output Layer |

Figure 8-3. *Multilayer perceptron*

Figure 8-3 shows an MLP with three nodes at the visible layer, four nodes at the hidden layer, and an output layer with only one possible outcome. Listing 8-12 imports the Keras library.

Listing 8-12. Import Libraries

```
from keras import Sequential, regularizers
from keras.layers import Dense, Dropout
from keras.wrappers.scikit_learn import KerasClassifier
```

After importing the Keras framework, we start establishing the structure of the neural network.

Architecture

Listing 8-13 builds up the logical structure of the neural network. There are eight input variables (High, Low, Open, Close, Volume, Adj Close, pct_change, log_ret, direction), and we intentionally set input_dim as 8 and apply the sigmoid function. The sigmoid function operates on a set of input values and generates output values that range between 0 and 1. We implement the ReLu function on the hidden layer and the output layer. Unlike the sigmoid function, the ReLu function limits the data between 0 and 1, and it processes the data until it invariably produces an optimal value. We typically use the adaptive moment estimation (Adam) optimizer over other optimizers because at most it generalizes data better than its predecessors, especially when dealing with a large dataset. It extends both from Adadelta (an optimizer that properly adjusts learning rates based on a moving window of the modified gradients) and RMSProp (estimates the notable difference between the gradients' current weights and preceding weights; thereafter, it estimates the square root of the obtained standard deviation). Adam is straightforward; it subtly alters the adaptive learning rate and implements the stochastic gradient descent method (a prevalent method for training neural networks faster by randomly selecting terms to estimate rather than estimating all terms). It adequately considers substantial variations in the loss function. In addition, it is not computationally demanding. In summary, it solves the problem of slow training (which we mostly experience with other optimizers when there are many variables or many observations in the data).

Listing 8-13. Architecture

```
def create_dnn_model1(optimizer="adam"):
    model1 = Sequential()
    model1.add(Dense(8, input_dim=8, activation="sigmoid"))
    model1.add(Dense(8, activation="relu"))
    model1.add(Dense(1, activation="relu"))
    model1.compile(loss="binary_crossentropy",
optimizer=optimizer, metrics=["accuracy"])
    return model1
```

Listing 8-14 wraps the architecture of the network using the KerasClassifier() method.

Listing 8-14. Wrap the Model

```
model1 = KerasClassifier(build_fn=create_dnn_model1)
```

Finalize the Model

Listing 8-15 trains the neural network across 64 epochs in 15 batches. An epoch represents a complete forward and backward pass, and a batch represents the number of samples that progressively increase throughout the network.

Listing 8-15. Finalize the Model

```
history1 = model1.fit(x_train, y_train, validation_data=(x_
val,y_val), batch_size=15, epochs=64)
history1
```

Listing 8-16 returns the key performance evaluation metrics (see Table 8-9).

Listing 8-16. Classification Report

```
y_predmodel1 = model1.predict(x_test)
creportmodel1 = pd.DataFrame(metrics.classification_report(y_
test,y_predmodel1, output_dict=True)).transpose()
creportmodel1
```

Table 8-9. *Classification Report*

	precision	recall	f1-score	support
0	0.468938	1.000000	0.638472	234.000000
1	0.000000	0.000000	0.000000	265.000000
accuracy	0.468938	0.468938	0.468938	0.468938
macro avg	0.234469	0.500000	0.319236	499.000000
weighted avg	0.219903	0.468938	0.299404	499.000000

Table 8-9 highlights that the neural network is less accurate and precise than the logistic classifier. It also shows that the data is imbalanced. You can further understand how skillful the classifier is by looking at the loss.

Training and Validation Loss Across Epochs

Loss represents a metric that establishes the difference between actual values and those predicted by the model. Listing 8-17 plots the training and validation loss across epochs to show how the neural network learns to differentiate between a downward and upward market in training and in cross-validation (see Figure 8-4).

Listing 8-17. Training and Validation Loss Across Epochs

```
plt.plot(history1.history["loss"],color="red",label="Training
Loss")
plt.plot(history1.history["val_loss"],color="green",label="Cro
ss-Validation Loss")
plt.xlabel("Epochs")
plt.ylabel("Loss")
plt.legend(loc=4)
plt.show()
```

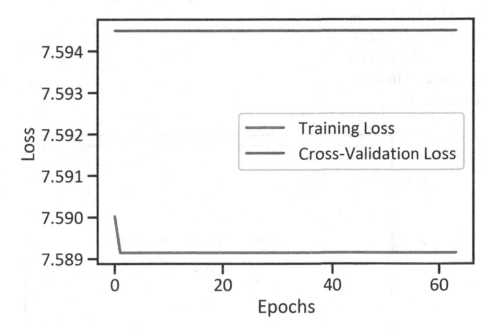

Figure 8-4. *Training and validation loss across epochs*

Figure 8-4 shows that at the first epoch, the training loss drops and remains until the 64th epoch (7.59). Meanwhile, the cross-validation loss remains constant across epochs (around 7.58).

Training and Validation Accuracy Across Epochs

Accuracy refers to how often a classifier correctly predicts classes. Listing 8-18 plots the training and validation accuracy to demonstrate how the neural network learns how to get the answers correct. (See Figure 8-5.)

Listing 8-18. Training and Validation Accuracy Across Epochs

```
plt.plot(history1.history["accuracy"],color="red",label="Traini
ng Accuracy")
plt.plot(history1.history["val_accuracy"],color="green",label="
Cross-Validation Accuracy")
plt.xlabel("Epochs")
plt.ylabel("Accuracy")
plt.legend(loc=4)
plt.show()
```

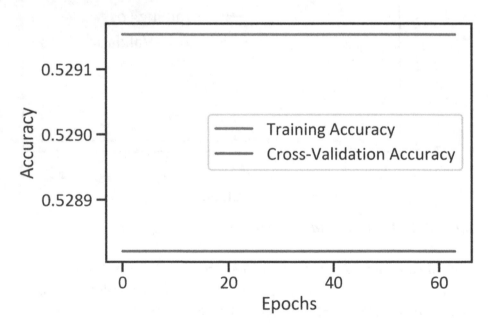

Figure 8-5. *Training and validation accuracy across epochs*

Figure 8-5 shows that both the training accuracy and the validation accuracy remain constant across epochs (with the training accuracy around 0.5291 and the cross-validation around 0.528).

Conclusions

This chapter introduced binary classification. It covered a way of designing, developing, and testing a machine learning model known as *logistic regression*, and it covered a neural network model known as the *MLP model* to solve a binary classification. It also showed metrics for classification model performance evaluation. After carefully reviewing the model's performance, we found that the independent variables are good predictors of the probability for an upward market and downward market.

CHAPTER 9

Investment Portfolio and Risk Analysis

This chapter properly concludes the book by covering a comprehensive framework for investment portfolio and risk analysis. Thus far, we have developed several machine learning models and deep learning models for robust investment management decision-making. Throughout the book, we alluded to investing in markets involving risk. In this chapter, we present the primitives of investment risk and performance analysis using the Pyfolio package. To install Pyfolio in the Python environment, we use `pip install pyfolio`, and in the conda environment, we use `conda install -c conda-forge pyfolio`. Before you install `pyfolio`, first install `theano`. To install `theano` in the conda environment, we use `conda install -c conda-forge theano`.

An investor typically invests in an asset expecting future financial returns. For instance, a company acquires plants and machinery to transform raw materials into products that can be sold to attain financial returns. There are ample asset classes that an investor can invest in. Common asset classes include stocks, bonds, equities, commodities, hedge funds, real estate, and retail investment trusts, among others. Each asset class has its own underlying characteristics and considerable value expressed in monetary terms. The value of an asset varies over time because of many factors. An investor must constantly monitor and improve an investment portfolio's performance.

© Tshepo Chris Nokeri 2021
T. C. Nokeri, *Implementing Machine Learning for Finance*,
https://doi.org/10.1007/978-1-4842-7110-0_9

Investment Risk Analysis

A clear understanding of the risks associated with an asset enables effective planning for unfavorable market conditions that may occur. By default, investors get exposed to risk when they take a position in the market, irrespective of the side of the position they take. When the market moves against an investor's position, they suffer losses. Managing losses is key to stable and viable investment strategies. Table 9-1 describes basic investment performance metrics.

Table 9-1. *Basic Investment Performance Metrics*

Metric	Description
Value at risk	Representing the minimum capital required to compensate for losses at a specified probability level
Drawback	Representing the rate at which an asset suffers losses
Volatility	Representing the extent at which an asset's price deviates away from the true mean value

Pyfolio in Action

This chapter convincingly demonstrates the Pyfolio package, a powerful package for investment risk and performance analysis. We can use this package with other complementary packages like Zipline and Quantopian to backtest an investment strategy. Backtesting represents considering how well an investment system manages risk and makes returns. We base it on the notion that past performance impacts future performance. In this chapter, we use the Pyfolio package as a stand-alone package. Before using it, we first extract the data. Listing 9-1 scrapes market data from Yahoo Finance by applying the get_data_yahoo() method. In this chapter, the

Amazon[1] performance is benchmarked alongside the Standard & Poor (S&P) 500 index.[2] Amazon is a US-based computer company giant whose stock is exchanged as an S&P 500 component. The S&P 500 is a stock index for gauging the performance of 500 companies listed in the United States.

Listing 9-1. Scraped Data

```
from pandas_datareader import data
import pyfolio as pf
ticker = 'AMZN'
start_day = '2010-10-01'
end_day = '2020-10-01'
amzn = data.get_data_yahoo(ticker, start_day, end_day)
spy = data.get_data_yahoo('SPY', start_day,end_day)
```

After web scraping, we perform necessary feature engineering tasks (see Listing 9-2).

Listing 9-2. Estimate the Returns

```
amzn = amzn["Adj Close"].pct_change()[1:]
spy = spy["Adj Close"].pct_change()[1:]
```

After estimating the daily returns, we use varying matrices to test the performance of the Amazon stock.

Performance Statistics

The `pyfolio` package enables us to comprehensively examine the fundamental performance of an investment strategy with some simple code. Listing 9-3 tabulates the Amazon stock performance from November 1, 2010, to November 2, 2020 (see Table 9-2).

---

[1]https://finance.yahoo.com/quote/AMZN
[2]https://finance.yahoo.com/quote/%5EGSPC/

Listing 9-3. Performance Results

```
pf.show_perf_stats(amzn, spy)
```

Table 9-2. *Performance Results*

Start date	2010-10-04
End date	2020-10-01
Total months	119

Backtest	
Annual return	35.6%
Cumulative returns	1995.7%
Annual volatility	31.6%
Sharpe ratio	1.12
Calmar ratio	1.04
Stability	0.97
Max drawdown	-34.1%
Omega ratio	1.23
Sortino ratio	1.71
Skew	0.42
Kurtosis	7.53
Tail ratio	1.07
Daily value at risk	-3.8%
Alpha	0.23
Beta	1.02

Table 9-2 highlights the risk that an investor bears when they exchange Amazon stocks. We use the SPY stock index as the benchmark in our analysis. It specifies that the annual rate of return is at 35.6 percent, and the cumulative return is at 19995.7 percent. Relating to risk, the daily value at risk is at 3.8 percent, and the maximum drawdown is at -34.1 percent. When observing investment performance findings, at most we are interested in studying three key ratios: Calmar ratio, Beta ratio, and Sharpe ratio. We equally devote our attention to alpha and beta. Underneath, we take you through how we measure these ratios. Table 9-3 provides a high-level overview of the key performance result mentioned.

Table 9-3. *Key Performance Results*

Metric	Description
Calmar ratio	Representing estimates of the uniform annual rate of return divided by the maximum drawdown
Beta ratio	Representing estimates of the difference between the anticipated return from investment and the anticipated return from the market over an exact risk-free return
Sharpe ratio	Representing estimates of the difference between a risk-free rate and portfolio returns
Alpha	Representing estimates of investment returns compared to a market key index
Beta	Representing estimates of volatility associated with an investment

In the following section, we study the underwater maximum drawback across time.

Drawback

Listing 9-4 applies the `plot_drawdown_underwater()` method to exhibit the rate at which an investment strategy suffered losses for 10 years (see Figure 9-1).

Listing 9-4. Portfolio Drawback

```
pf.plot_drawdown_underwater(amzn)
plt.show()
```

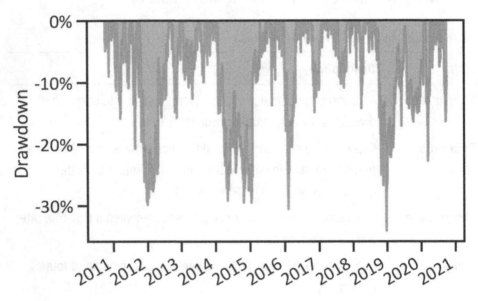

Figure 9-1. *Underwater drawback plot*

Figure 9-1 postulates that the drawdown of the Amazon stock was between 0 percent and -30 percent from November 1, 2010, and November 2, 2020. In 2019, they vicariously experienced the uppermost drawdown; nevertheless, in the subsequent year, they promptly recovered losses.

Rate of Returns

The most convenient way of determining whether an investment is attractive involves studying the rate of returns. Rate of return shows the period's beginning price and the period's end price.

Annual Rate of Return

The annual rate of return represents the rate at which an asset yield returns annually. We study returns over time to make forecasts. Listing 9-5 applies the plot_annual_returns() method to plot the annual rate of returns of Amazon stocks over 10 years (see Figure 9-2).

Listing 9-5. Annual Rate of Returns

```
pf.plot_annual_returns(amzn)
plt.show()
```

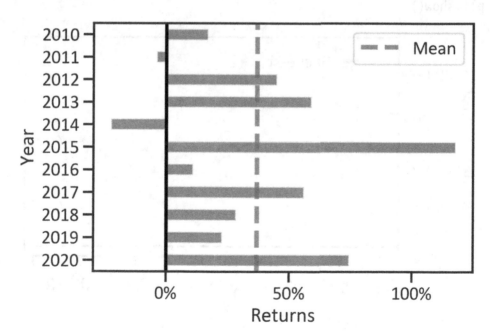

Figure 9-2. *Amazon annual rate of return*

Figure 9-2 shows that Amazon stocks produced an instable annual rate of return from November 1, 2010, to November 2, 2020. There are changes in annual returns. In 2014, the portfolio had the worst performance. However, in the following year, the annual returns exceeded 100 percent.

Rolling Returns

Listing 9-6 returns backtested cumulative returns over 10 years (see Figure 9-3). Unlike the annual rate of returns, we weigh the cumulative returns. Pyfolio estimates the cumulative returns across time by estimating the notable differences between current returns and preceding returns (representing the total amount gained from an investment) over the cost of the stock.

Listing 9-6. Rolling Returns

```
pf.plot_rolling_returns(amzn)
plt.show()
```

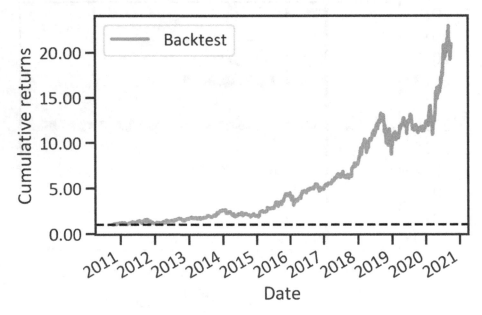

Figure 9-3. *Rolling rate of returns*

Figure 9-3 shows clear small increments in cumulative returns from November 1, 2020. In 2016, an upward trend started gaining momentum.

Monthly Rate of Returns

The monthly rate of returns represents the rate at which an asset yield returns monthly. We review rates of returns monthly to identify historical behavior and cautiously make short-term to midterm forecasts. Listing 9-7 applies the plot_monthly_returns_heatmap() method to plot a heatmap of Amazon stocks' monthly rate of return over 10 years (see Figure 9-4).

Listing 9-7. Monthly Rate of Return Heatmap

```
pf.plot_monthly_returns_heatmap(amzn)
plt.show()
```

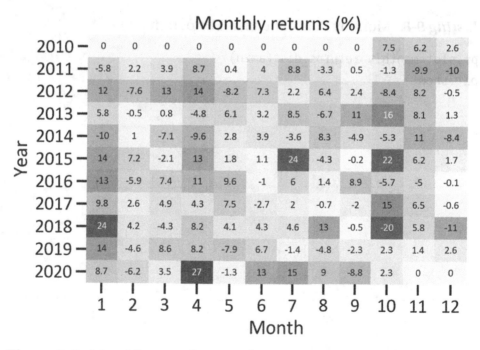

Figure 9-4. *Monthly rate of returns heatmap*

Figure 9-4 shows that Amazon stock had unstable monthly rates of return from November 1, 2010, to November 2, 2020. In the first nine months of 2010, the rate is represented as zero, which is the same as in the last two months of 2020. This is because we extracted data from November 1, 2010, to November 2, 2020. Figure 9-4 indicates returns were at their peak (at 27 percent) in April 2020. In October 2018, the stock experienced its worst performance; the monthly rate of return was -20 percent. Also, from 2017 to 2020 the stock shows positive returns, which was the same as in the fourth month of the year. For clarity, we carefully examined the distribution of the data to summarize the monthly rate of returns. The most widespread distribution is the normal distribution. Data follows a normal distribution when actual values saturate around the true mean value. Listing 9-8 applies the plot_monthly_hist() method to plot a histogram of the monthly rate of returns of the Amazon stock (see Figure 9-5).

Listing 9-8. Monthly Rate of Returns Histogram

```
pf.plot_monthly_returns_dist(amzn)
plt.show()
```

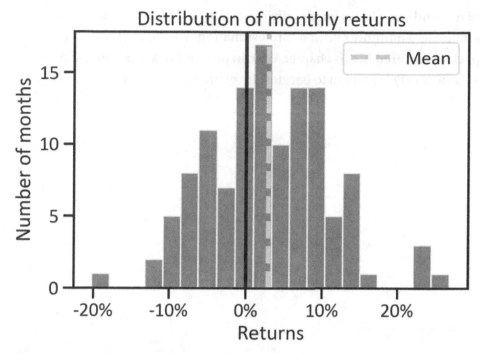

Figure 9-5. *Monthly rate of returns histogram*

The actual monthly rate of returns is scattered around the mean value. We can sufficiently summarize the data using the sample mean value. From November 1, 2010, to November 2, 2020, on average, the Amazon stock exhibited a long-run upward trend.

Conclusions

This chapter concludes the book in which we introduced ways to remedy financial problems, especially investment management problems, using machine learning and deep learning. It covered several ways for objectively analyzing the performance of an investment portfolio using a Python package known as Pyfolio. To begin with, it discussed the concept of risk and techniques for identifying risk exposure. Last, it covered an annual and monthly rate of return estimation. We are not limited to the package covered in this chapter. We can use the packages alongside Zipline and Quantopian to backtest investment strategies.

Index

A

Activation function, 52

Adaptive movement estimation (Adam), 52

Additive model
data preprocessing, 45
forecast, 47
model, 46
seasonal decomposition, 48, 50

Area under the curve (AUC), 155

Augmented Dickey-Fuller (ADF) test, 21

Australian Securities and Investments Commission (ASIC), 7

Autoregressive integrated moving average (ARIMA) model, 21
definition, 35
develop, 37
forecasting, 38–40
hyperparameters, 36, 37

B

Backtesting, 19

Bear trend, 74

Binary classification, 17

Brokerage, 6, 7

Bullish trend, 74

C

Classification method
accuracy across epochs, training/validation, 164
architecture, 161
data preprocessing, 150
dataset, 145, 147, 149
finalize method, 161, 162
loss across epochs, 162, 163
scrap data, 144

Cluster analysis, 18, 91

Contracts for difference (CFD), 9, 92

Correlation method
covariance, 105
definition, 103
Eigen matrix, 110, 112
pairwise scatter
plot, 106–108
Pearson, 104

D

Deep learning model, 51, 167

Desk dealing (DD) brokers, 7

Dickey-Fuller (ADF) test, 27, 73

Dimension reduction, 18

Printed in the United States
by Baker & Taylor Publisher Services